The Founders' Folly

The Predictable Patterns That Undermine Technology Startups

by Gary Smart

For those who sensed something was wrong
long before they could name it.

The Founders' Folly

The Predictable Patterns That Undermine Technology Startups

This book is a work of non-fiction grounded in the author's lived experience. Some identifying details have been altered or generalised, and some situations have been compressed or lightly blended at the edges, to protect people and contexts. The book is concerned with recurring structural patterns rather than providing a definitive account of any individual, company, or project.

The material in this book is provided for general informational and reflective purposes. It does not constitute business, financial, legal, investment, or other professional advice. Readers should exercise their own judgement and seek qualified advice appropriate to their circumstances.

First edition 2026

Independently published by Gary Smart

1.1-20260912

Contents

Acknowledgements

This book was shaped by conversations, experiences, and moments shared with many people over many years.

I'm grateful to the founders and leaders — technical, operational, and strategic — who carried responsibility, to the engineers and developers who shaped the work, and to the advisors and teams who supported the journey — often under pressure, often in uncertainty, and always with good intent.

While this book reflects patterns rather than individual stories, it exists because of the trust and openness of those environments.

I'm also grateful to the friends, colleagues, and associates I have worked with across the past three decades who I have been able to reconnect with, and who have shared their experience and reflections on our ventures together, as well as stories from other adventures that have helped validate and enrich the patterns explored in this book.

I'm also grateful to the early readers who gave their time and attention to drafts of this work, and to those who challenged my thinking and sharpened my perspective.

Thank you to those who supported the long, quiet process of turning experience into something that could be named.

Preface

This book is built from real startup experiences.

The material in this book is not built from hypothetical examples, borrowed case studies, or abstract management ideas. It comes from situations I have lived through across three decades of technology startups, projects, product teams, delivery environments, funding conditions, and changing technical waves.

The majority of the stories and examples in this book are drawn closely from specific ventures in which I have been deeply involved. Others are lightly anonymised, compressed, or blended at the edges to protect people and contexts. But if a situation appears in this book, it reflects something that genuinely happened. The exact scene may belong to one venture, but the mechanism underneath it is there because it repeated across different startup projects.

This book is not trying to explain isolated failures or retell the story of a single company. It is trying to name recurring structural patterns that appear across different startups, teams, and technologies, often under very different surface conditions.

As I have written it, I have also reconnected with friends, colleagues, and former founders and leaders from across that same period of work. Their reflections, corrections, and adjacent stories have helped validate, sharpen, and deepen what is written here. The patterns described in this book are grounded in my own experience and judgement, but they have been strengthened by being checked against other people who lived through related versions of the same dynamics.

Every technology startup begins with a necessary lie about certainty.

It has to. If founders were fully honest about the odds, the uncertainty, the trade-offs, and the personal cost, most companies would never be started. Confidence is not optional; it is the fuel. Belief is not a flaw — it is a prerequisite.

And yet I have watched that same belief harden over time into something more dangerous. It can be summarised simply:

Every founder believes their startup is the exception to history.

- "We are different."
- "We have learned from everyone else's mistakes."
- "We hire the smartest people."
- "That is what we need for now. We will fix it later."

I have heard these phrases spoken by thoughtful, intelligent, well-intentioned founders, investors, and senior leaders. I have also heard them spoken moments before companies entered irreversible strain or irreversible decline.

This book is not written to assign blame, nor is it a retrospective catalogue of failure. It is written from inside the room, while belief still feels reasonable and while the fate of the company is still being shaped by everyday decisions.

The most dangerous dynamics in startups rarely announce themselves as mistakes. They emerge slowly through decisions that feel justified, necessary, and even responsible at the time. When they are recognised clearly, behaviour has adapted, systems have adjusted, and the range of possible futures has narrowed.

What follows is not theory, but a way of seeing. Many startup books provide roadmaps for how a company should progress from idea to scale. This book looks instead at the weather systems that form around those journeys.

If you are founding a technology startup, leading one, investing in one, or carrying responsibility for the technical or structural decisions that will shape its trajectory, this book is for you.

If you are just starting out in your career, entering a startup, or thinking about founding a technology startup, I hope this book offers something harder to get any other way: hindsight in the form of foresight. Many of the technical, business, and personal pressures you are likely to encounter are visible in these pages. You can choose to recognise them early or encounter them later — but they are not going away.

The first task is not to redesign the system, but to recognise what it is communicating. To see pressure as pressure. To see drift as drift. To notice when momentum and distortion begin to blur.

The basic point is this: these patterns are not rare. They are just rarely named clearly while they are still underway.

What follows begins with one such story.

Part I

Reading the Weather

"Beliefs justify behaviours. Behaviours create blind spots. Blind spots delay reality."

Chapter 1
Introduction

The room was full.

Around twenty developers. Project managers. Senior leaders.

Years of effort sitting quietly in chairs, laptops closed.

It was the year 2000. The world had survived the Y2K bug, and the dot-com boom was in full swing, and this was *the* startup to be working on.

Money wasn't tight. Expectations were high. Tens of millions of dollars had been spent.

At the front of the room stood the most senior representative from the business.

Always calm. Always smiling. Not fake — just practised. He had a way of delivering difficult news without ever sounding difficult himself.

Today, though, he was congratulating us.

He thanked the team for pushing through challenges. He spoke about progress. About momentum. About how pleased he was with where the project was heading.

And then it happened.

A look passed between those of us closest to the work.

No words. Just shifting eyes. A tightening in the chest.

Nothing was said, but the same quiet question moved across the room:

He doesn't know, does he?

You could feel the change immediately. The air thickened. Smiles tightened. Nods became automatic. The applause came — polite, but uneven. Hollow.

The man at the front noticed.

His smile stayed, but his posture shifted. He slowed down. Paused longer than before. This time, he wasn't reading reports or metrics. He was reading faces.

The meeting ended early.

No confrontation. No explanation. Just an unsettled quiet as people packed up and walked out.

Within a week, the project was terminated.

Not paused. Not redirected. Not "under review."

It was simply over.

Years of work disappeared without ceremony.

For those closest to the work, this hadn't come out of nowhere.

We had been feeling the strain for months — in the performance issues, the growing shortcuts, the constant sense of propping things up just to keep them moving. Technical risks were raised, discussed, documented, and escalated as best they could be. But somewhere along the way, the full weight of what was happening stopped travelling upward.

Technical debt accumulated. So did a deeper exhaustion and human toll that never appeared in plans, metrics, or status updates. Morale thinned. Confidence became brittle. People were working in worsening conditions long before anyone at the top realised the weather had changed.

For the man championing the project inside the organisation — the one carrying real professional, political, and reputational stakes — it arrived all at once.

There was no gradual realisation.

No opportunity to adjust course.

One moment he believed the skies were clear.

The next, he was standing in the middle of a storm that had been forming for months.

That was my first real lesson in startups.

And it wouldn't be the last time I watched a serious, well-funded company sail confidently forward, convinced the conditions were stable — while those closer to the work had already begun to feel the pressure dropping and the winds changing.

The storm had been building for a long time.

Only some people could see it.

I didn't understand what I had just witnessed at the time.

I knew the project was over. I knew people had lost their work, their momentum, and months — even years — of their lives. But I didn't yet have language for why it had ended the way it did, or why the collapse felt so sudden to some and so inevitable to others.

What stayed with me wasn't the outcome itself.

It was the timing.

The people closest to the work had felt the strain building long before leadership realised how unstable things had become. The person with the highest stakes discovered the truth all at once. Between those two realities sat layers of process, optimism, translation, and silence.

At the time, I treated it as an unfortunate anomaly rather than a warning. A product of its era. Dot-com excess. Bad luck.

But over the years, as I moved through more technology startups, a more uncomfortable recognition set in.

The details changed. The technologies changed. The people changed.

But the experience these companies moved through did not. And when the high-energy honeymoon period gave way to reality and challenges emerged, the *shape* of the struggle — and in the worst cases, failure — remained strikingly similar.

In that early period, enthusiasm can feel like evidence. Culture can feel like structure. A room full of smart people can start to look like proof that the startup will solve whatever reality puts in front of it.

Sometimes that energy is valuable fuel.

The mistake begins when it is mistaken for validation.

Different companies. Different leadership. Different levels of funding. The same quiet signals. The same confident narratives. The same moment where certainty vanished faster than anyone expected.

It became clear over time that what I had witnessed in that room was not a one-off collapse.

It was something much more common.

A pattern that repeats inside ambitious companies — long before anyone has language to describe it.

Chapter 2
Interpreting the Signals

Many of the most influential startup books focus on stages, execution, and growth mechanics. They emphasise experimentation, positioning, scaling, and leadership under pressure. Together, they provide practical frameworks for product–market fit, fundraising, hiring, and growth. They teach what to do at each stage and how to progress with structure and discipline. In many ways, they are roadmaps — clear guides that move founders from milestone to milestone.

The Founders' Folly sits alongside those roadmaps, but it asks a different question. Instead of mapping stages, it studies the weather.

A roadmap tells you where you are going. Weather tells you what conditions you are travelling through.

You can follow a precise map assuming clear skies and still drive straight into a storm.

Founders do not ignore the possibility of bad weather — they know the statistics. But most assume they will see the storm forming, read the instruments correctly, and adjust in time. They trust the dashboard — the metrics, the runway, the growth curves — to warn them early enough.

Those dashboards measure progress. They do not always measure conditions.

One reason this becomes so dangerous is that startups often get better at producing the appearance of progress than at improving the conditions that make progress real.

A convincing demo. A polished prototype. A rising metric. A successful investor conversation. A roadmap full of movement. None of these are meaningless. But neither are they the same as structural health. In many companies, what begins as genuine momentum gradually hardens into something more deceptive: the appearance of progress. The organisation learns to generate visible proof that it is advancing, while the underlying system becomes harder to trust, harder to change, or more dependent on controlled conditions than anyone wants to admit.

That distinction is important because the appearance of progress does not feel deceptive while it is happening. It feels encouraging. It gives teams, founders, and investors enough evidence to keep believing the company is close, even when the underlying system has stopped becoming more trustworthy.

But there is another complication that most founders never see at first.

Inside many startups, the signals are not coming from a single dashboard at all. They are coming from two very different ones.

Leadership tends to read the external signals of momentum — investor conversations, successful demos, partnership interest, and visible product progress. Engineers, meanwhile, read the internal signals of structural health — system stability, data integrity, architectural strain, and the growing effort required to keep things running.

I have seen this divide appear in small, ordinary ways. A leadership update describes a product as close because the demo works, the pitch is landing, and the next customer conversation is booked. In the same week, the people closest to the system may be discussing hard-coded paths, unreliable data, fragile deployment steps, or the manual work required to keep the demo route clean. Both groups are looking at real evidence. They are not reading the same condition.

Later in this book we will look more closely at what happens when those dashboards diverge.

Most roadmaps assume stable environments: rational actors, clean execution, aligned incentives, and clear visibility. In reality, startups rarely operate in blue-sky conditions for long.

Storm systems form gradually inside organisations.

Pressure builds. Visibility narrows. Tension rises. Signals become harder to interpret. Small course corrections go unnoticed while momentum continues forward. From the outside, progress appears intact. From the inside, something feels different.

A company can execute each phase correctly and still feel culture tightening, trade-offs compounding, and judgement distorting under invisible pressure.

Stages describe where a company is.

Patterns describe the conditions it is operating in.

This book focuses on those conditions.

It reveals the predictable patterns that shape technology startups regardless of stage. Across this book, those patterns trace a recurring arc: allusions, accumulation, human cost, and reckoning.

It also clarifies why not every failure begins the same way. Sometimes the product never really works. Sometimes it works well enough that customers genuinely like it and deployment holds, yet the company still fails for reasons the technology alone cannot explain. Pattern awareness helps you read both kinds of story without mistaking technical validity for structural health.

When you recognise them early, execution sharpens. You see narrative lock-in forming. You notice urgency masking fear. You recognise when compromises are stacking. You can intervene before accumulation becomes reckoning.

The challenge is not only to notice the signals, but to understand why different people inside the same company experience them so differently.

To interpret those signals clearly, we must first understand how they distort as they move upward through the organisation.

Chapter 3
Altitude Distortion

One reason signals are so difficult to interpret inside startups is that people rarely see the same system.

They see it from different altitudes.

Engineers working directly with the system experience its behaviour in detail. They see the edge cases, the workarounds, the fragile components that require constant attention. Small inconsistencies are visible because they live inside them every day.

Product managers and team leads tend to see a slightly different picture. They observe how features evolve, how priorities shift, and how work moves through the organisation. Their view connects technical reality with delivery and coordination.

Executives and investors usually experience the system through summaries — dashboards, milestones, metrics, and presentations. At that level the system appears more abstract. Complexity collapses into trends, progress reports, and narratives about direction.

At that altitude, direction itself can also become easier to move. A senior figure can reshape priorities, redefine the target, or push a new interpretation of the product without living inside the daily implementation cost of that change.

This creates a kind of impedance mismatch. From commercial altitude, flexibility looks valuable: change the pitch, shift the target, respond to the market. Closer to the work, that same movement can remove the stable ground needed to design, build, and integrate anything coherently.

None of these perspectives are wrong. They are different vantage points within the same system.

But altitude is not only a natural feature of organisational life. It is also shaped by design.

Who has direct access to the founder. Who must work through layers. Who is rewarded for confidence. Who is expected to translate uncertainty into reassurance. Who gets to slow a decision down, and who is expected to make it work after the fact. These are not merely interpersonal details. They affect how clearly reality can travel through the company.

Two startups may experience the same technical strain and respond very differently depending on how power, trust, and decision rights were arranged early on. In one, concerns can travel upward with enough force to reshape the plan. In another, they arrive as softened summaries. The organisation has learned that momentum outranks friction, that some truths are easier to carry privately than to escalate cleanly, and that certain versions of reality travel more safely than others.

As altitude increases, detail compresses. The system becomes easier to describe, but harder to feel. Signals that are obvious close to the work often travel upward as softened summaries, translated updates, or simplified explanations.

From the ground, you feel the wind changing.

From higher altitude, the sky can still look clear.

That was the split in the room that opened this book. From the front, the project still looked like a difficult effort moving forward. From the seats closest to the work, the instability had become part of daily life. The same startup was being read at two different altitudes.

Most of the time this difference is harmless. Organisations depend on abstraction. No leader can hold the full detail of a complex system in their head, and no engineer can see the entire strategic landscape at once.

But under pressure, altitude can distort interpretation.

Small technical compromises that feel temporary to those close to the work may appear invisible at higher levels. Workarounds that engineers experience as structural friction may look like ordinary iteration when they appear in status updates.

The same distortion can affect strategy. Repeated direction changes may feel like responsiveness from above while landing below as confusion, rework, and a moving target the team no longer knows how to build toward cleanly.

The signals are still present.

They simply weaken as they travel upward.

But the distortion is not always passive. Sometimes the people closest to the work do raise the harder questions, only to be routed around, treated as friction, or pushed back down because they threaten the preferred story. In those cases the problem is not only that truth softens on the way up. It is that the organisation learns which truths are unwelcome before they ever get there.

When the distortion becomes visible at higher altitude, the system underneath may already have changed significantly.

This difference in perspective does not require anyone to ignore reality. It emerges naturally from the way organisations translate information through layers of responsibility.

Many of the patterns in this book begin at ground level long before they become visible higher up. To interpret them clearly, we need to look beneath the signal itself to the structure producing it.

Chapter 4
The Structure Beneath

This book is structured as a lens on startup conditions rather than a startup roadmap.

Many startup frameworks describe stages: idea, product–market fit, scale, maturity. Those models are useful, but they focus on where a company is supposed to go.

This book looks at something different.

It separates the overlapping forces that shape what a company is actually experiencing while it is moving through those stages.

In fast-moving startups, many dynamics happen at once: optimism, pressure, speed, compromise, exhaustion, belief, and uncertainty. Because they appear together, they are rarely examined clearly while they are forming. They blur into what feels like "the reality of startup life."

The structure of this book slows that blur down.

Instead of treating startup struggles as isolated problems or management mistakes, it separates the forces that interact beneath the surface. By isolating them, the patterns become easier to see.

You can think of this structure like reading a weather system.

From a distance, a storm appears as a single event. But meteorologists separate what is happening into different forces — pressure, temperature, wind, and movement — because understanding each component makes the larger system visible.

This book does the same thing for startups.

After this introductory section, the pattern arc of the book is organised around four recurring conditions that tend to appear across technology startups under pressure. They are not business stages like idea, product-market fit, scale, or maturity. They are structural conditions that often overlap inside the same company, even though earlier distortions tend to create the conditions for later consequences.

In practice, these conditions may overlap. A company may already be carrying elements of accumulation while still speaking in the language of allusions. Human cost may be visible to some people long before reckoning is visible to everyone. But the arc still matters, because earlier conditions create later consequences.

Taken together, these four conditions trace a recurring arc. Each one acts as a lens onto a different layer of the same underlying pattern.

Part II – The Allusions

This section explores the narratives that give startups their initial momentum. Optimism, speed, confidence, and talent are not mistakes — they are necessary conditions. But the stories organisations tell themselves can gradually become more persuasive than evidence. Over time, those narratives shape interpretation itself, overriding history, process, and dissent long before anything appears obviously wrong.

Part III – The Accumulation

Here the focus shifts to what builds under pressure. Short-term compromises, exceptions, and "we'll fix it later" decisions stack. Technical debt grows, process shortcuts appear, and trade-offs compound. Individually, each decision is reasonable. Together, they gradually reshape the system that is actually being built.

Part IV – The Human Cost

This section examines what happens to the people closest to the work. As strain increases, communication frays. Signals stop travelling upward. Behaviour adapts in ways that allow the system to keep functioning even as the underlying pressure grows. Burnout, disengagement, silence, and emotional withdrawal appear not as personal weaknesses but as predictable responses to sustained structural strain.

Part V – The Reckoning

This final pattern section examines what happens when deferral ends and accumulated trade-offs become visible. What once felt flexible can no longer be postponed. Earlier decisions assert themselves. Constraints surface not because of a single mistake, but because the margin for avoidance has gradually disappeared. The reckoning often feels sudden — not because it arrived without warning, but because its formation was rarely acknowledged while it was unfolding.

Taken together, these four parts describe a pattern arc that appears repeatedly across technology startups.

Allusions shape perception. Accumulation reshapes reality. Human cost absorbs the strain. Reckoning reveals what has been forming.

To recognise these dynamics while they are still forming, we must first be precise about what they are.

Chapter 5
What I Mean by a Pattern

The central idea in this book is pattern.

By patterns, I mean recurring dynamics that appear across startups, leadership teams, and individuals — regardless of industry, technology stack, funding model, or market timing. In practice, these dynamics rarely belong to a single role or decision. They emerge from the interaction of many choices made under pressure.

Throughout this book, when I use the word *founder*, I'm using it in a broad and practical sense — not only the person who started the company, but anyone whose decisions materially shape its direction. That includes technical leaders, executives, investors with influence, and senior decision-makers whose choices affect the trajectory of the organisation.

The word *folly* matters too.

I am not using it to mean stupidity, recklessness, or simple misjudgement. I use it to describe what happens when belief, pressure, ambition, and altitude combine to delay visibility. Important things can be happening inside a startup long before the people with the most influence, or the most reputational stake, can see them clearly.

That is the sense in which I use the title *The Founders' Folly*.

Patterns are not bugs in execution or signs of poor intelligence. In fact, they tend to appear most reliably in organisations that are smart, ambitious, and moving quickly.

A pattern is not a single mistake. It is a shape.

It appears in the language people use, the shortcuts that feel justified, the conversations that stop happening, and the decisions that are deferred just long enough to feel reasonable. While you are inside a pattern, nothing feels obviously wrong. The decisions feel sensible. The explanations feel convincing. The progress often appears real.

Patterns usually become obvious only later — when someone looks back and says:

"I've seen this before."

Founders and senior leaders recognise patterns instinctively, even if they do not always name them. You hear it in phrases like:

"This feels familiar."

"I've seen this movie before."

"Something's off, but I can't quite point to it."

That moment — when something feels familiar before it is fully understood — is pattern recognition beginning to surface.

What makes these patterns dangerous is not that they are irrational. Most of them are locally sensible. Each decision, on its own, can be explained. Each trade-off feels temporary. Each exception feels earned.

The chapters that follow focus on the book's keystone patterns: the primary structural patterns that help explain many smaller behaviours. A startup may experience dozens of local symptoms, such as hiring too quickly, resetting the stack, letting demos become product claims, or treating silence as alignment. Those symptoms matter, but they often gather around a smaller number of larger patterns.

The problem is that patterns compound.

Small compromises stack on top of one another under pressure. Beliefs reinforce behaviour. Behaviour creates blind spots. Blind spots delay reality. These dynamics reshape the system over time — long before metrics, incidents, or financial results make anything undeniable.

In environments where tools such as AI accelerate output and shorten feedback loops, this compounding can happen faster than shared understanding can keep up.

This is where startups accumulate invisible cost.

Once consequences are visible, reversal may be expensive — or no longer fully recoverable. Technical debt that cannot realistically be unwound. Teams that were once tightly aligned becoming depleted, fractured, or lost through churn. Momentum that appears intact on the surface while coherence underneath has eroded.

Many of the patterns described in this book do not emerge from bad intent or poor execution. They emerge from predictable tensions present in almost every fast-moving startup, often at the intersection of:

- speed and fear
- incentives and identity
- optimism and power
- responsibility and distance from the work

These are human patterns, not technical ones.

Technology changes.

Human dynamics do not.

As you read, some of these patterns may feel immediately familiar. Others may not — especially if this is your first time inside a startup moving at speed. That is intentional. These dynamics rarely feel dramatic when they first appear. They surface as reasonable decisions, sensible compromises, or short-term wins.

The value of naming them is not just recognition in hindsight. It is awareness in advance.

If months or years from now you notice one of these patterns forming — or recognise the signals that usually accompany it — that is the moment to pause. Not because failure is inevitable, but because the costs being accumulated are often far larger than they appear at the time, and far harder to reverse once they become visible.

The pattern chapters therefore return to a similar set of questions: how the pattern forms, what it looks like early, what gets misread, what cost it hides, and what recognition might make possible before the pattern becomes more expensive.

There is one more thing worth naming upfront: these patterns rarely appear in isolation. They reinforce one another.

Beliefs justify behaviours. Behaviours create blind spots. Blind spots delay reality.

An internal logic forms over time — one that allows intelligent people to feel increasingly confident even as options narrow.

I will unpack that structure gradually as the book progresses. But keep this in mind as you read:

If several of these patterns feel familiar at once, it is rarely coincidence. It is usually a signal.

Chapter 6
Seeing Without Distortion

Seeing startups clearly is harder than it first appears.

If you are trying to build something ambitious in a startup, you are not only building code, product, or strategy. You are also building belief.

It is part of what makes the system possible. Founders need it. Early staff need it. Investors need it. Customers need it as well. Without some confidence about what might be possible, most startups would never get out of the starting blocks.

The difficulty begins when belief stops being fuel and becomes a filter.

It can change over time what a company notices, excuses, delays, and no longer feels able to question. From the inside, this rarely feels irrational. It feels necessary. That is part of the danger.

The patterns described here are recurring distortions in how signals are interpreted inside real startups, often long before anything looks obviously wrong. They surface quietly — in language, in meetings, in incentives, and in the way decisions are explained rather than examined, subtly reshaping what the organisation believes it is seeing.

Altitude matters here because altitude changes what a person is exposed to. The higher up you are in a company, the more mediated your view becomes. You receive summaries rather than texture. Polished explanations rather than raw uncertainty. Confidence-shaped language from intelligent people who are also trying to keep the work feeling coherent enough to move.

Distortion does not require deception. It can form naturally as information travels through layers of pressure, translation, and hope.

The urge to look for answers often arrives before signal interpretation has finished. Clarity is slower, and often more unsettling. That discomfort is part of the work.

Patterns are not problems to be solved in isolation. They are signals about how other signals are being filtered, amplified, or ignored inside the system. When organisations misread those signals, they often respond too quickly or too mechanically — creating new problems while leaving the original distortion intact.

A more useful stance is to slow interpretation down: to notice signals before reacting to them, to recognise familiar dynamics without immediately trying to correct them, and to allow the signal to clarify before deciding what it means.

Slowing interpretation down also means staying willing to revise the explanation you form. Recognising a familiar pattern should sharpen inquiry, not settle it.

Many of the patterns described here will feel reasonable. Some will even feel necessary. That does not mean they are harmless; it usually means they have become normal.

As you read, there will be a natural urge to ask:

"What should I do about this?"

But in the early stages of pattern recognition, different questions tend to matter more:

Where have I seen this? How long has this felt normal? What explanations do we use to keep this in place? What would it cost to acknowledge this honestly?

Failure is not inevitable. The point is that you are navigating complexity under pressure, and that many of the most costly outcomes in startups follow recognisable patterns. They repeat across different companies, founders, technologies, and markets because intelligent people under pressure learn to interpret signals in familiar ways.

Seeing clearly is not a weakness — it is the prerequisite for any meaningful response.

And in fast-moving environments — especially those shaped by AI-assisted development — the temptation to respond before seeing has finished becomes stronger, while the margin for correction becomes thinner.

The patterns in this book are separated rather than collapsed into one general warning so they can be seen clearly before they are judged or acted upon.

Later in the book, we will return to what it looks like when signal recognition surfaces in real time, and how leaders can stay oriented without rushing too quickly into action.

Chapter 7
AI, Speed, and Structural Acceleration

This book was shaped by patterns observed across decades of software development and startup life, many of them formed long before artificial intelligence became a daily tool in product teams.

That context is worth naming.

AI has radically changed how software is written, how quickly ideas move, and who can participate in building systems. It is the most significant shift in software creation since the rise of the internet itself. Today, it is entirely possible to build products, platforms, and businesses at a speed that would have been unthinkable even a few years ago.

What AI has not changed are the deeper, recurring forces that shape how startups behave over time, regardless of whether they ultimately succeed or fail.

Those forces existed long before AI. They exist now. And they will remain long after the tools themselves change.

The patterns explored in this book, organised into four major sections that reflect recurring phases of pressure — allusions, accumulation, human cost, and reckoning — do not originate in technology. They originate in human behaviour: how decisions are made under pressure, how responsibility is deferred, how complexity is rationalised, and how short-term progress is mistaken for long-term stability.

AI does not remove these forces. It amplifies them by reducing friction and compressing time.

But acceleration is not evenly distributed across the startup. AI can compress the time required to build, demonstrate, and extend a product. It does not compress the time required for customers to trust it, organisations to procure it, regulators to approve it, teams to understand it, or people to recover around it. When one clock accelerates and the others do not, the company can consume runway, confidence, and human margin faster without bringing the slower evidence any closer.

Ideas move faster. Prototypes appear sooner. Systems can look production-ready long before they are truly understood. Allusions form faster, accumulation happens sooner, and the gap between what works and what is understood widens more quickly. Systems can appear to approach production scale long before the underlying structure, behaviours, or shared understanding required for a true production system are actually in place.

In some teams, this creates a new kind of dependency: products that can only be extended, repaired, or reasoned about through the same tools that generated them. Human understanding becomes optional — until it suddenly becomes critical.

The reckoning still comes, but it arrives earlier, sometimes repeatedly.

This book focuses on what remains constant beneath every technological shift: the patterns that emerge when speed outpaces sense-making, when responsibility is abstracted away, and when progress is measured without ownership.

If anything, AI makes these patterns easier to see — if we are willing to look.

Part II

The Allusions

"Startups rarely unravel in moments that feel dangerous. They drift apart in moments that feel perfectly reasonable."

Chapter 8
When the Story Takes Hold

Startups rarely unravel in moments that feel dangerous. They drift apart in moments that feel perfectly reasonable.

There is no argument. No alarm. No obvious mistake. Just a distinction made under pressure, accepted without resistance, and quietly folded into the way work proceeds.

Moments like this rarely register as decisions. They feel like clarity. Like focus. Like maturity.

And yet it is often here — in these small, sensible distinctions — that the story a startup tells itself takes shape.

Every startup begins with a story.

Sometimes that story is explicit, written into pitch decks, strategy documents, and all-hands meetings. More often it is implicit. It lives in the language people use, the shortcuts they justify, the risks they normalise, and the assumptions nobody feels the need to question.

That story becomes the lens through which everything else is interpreted.

This part of the book examines those early distortions.

Not obvious breakdowns. Not post-mortems. But faint signals that appear while everything still seems to be working. Moments that, in hindsight, feel important — but at the time felt harmless, even reasonable.

Before going further, it is worth being precise about the word *allusion*. An allusion, with an A, is a reference that quietly points the mind toward a familiar story. It does not announce itself as false. It simply gives the mind a direction to complete.

An illusion, with an I, is different. It is what happens when the mind mistakes that completed picture for reality. The distinction is important in this section. The earliest signals are allusions: hints that point the organisation toward a story it already wants to believe.

Sometimes, when those signals become convincing enough, the allusion hardens into an illusion.

An allusion is not a warning siren.

It is a hint. A shadow. A subtle shift in tone.

A shift from *"we're learning"* to *"we'll clean that up after the next release."*

From *"this matters"* to *"this is good enough to ship."*

From *"that's a design issue"* to *"that's an operations problem now."*

Signals like these are rarely ignored through carelessness. They pass unnoticed because the environment rewards momentum, optimism, and belief. Speed masks fragility. Confidence overrides doubt. Temporary success stands in for truth.

In the chapters that follow, those allusions appear in different forms: as a story that stops changing, as visible progress that feels like proof, and as confidence borrowed from people, partners, tools, or credibility earned somewhere else.

In weather terms, these are the early pressure changes — the quiet indicators that conditions are shifting long before a storm becomes visible.

What follows is not a judgement of decisions made under pressure.

It is an attempt to hold still long enough to notice the shape of things forming beneath the surface — before they harden, before they demand explanation, and before they acquire consequences.

They are not alarms. They are the subtle references that point the mind toward a story taking hold.

They are allusions.

Chapter 9
Narrative Lock-In

When the company becomes trapped inside its own story.

Every startup tells itself stories. It has to. A story gives the team something to build toward, gives investors something to understand, and gives the market a reason to pay attention before all the evidence is in. At the beginning, that story is not the problem. It is part of how people move through uncertainty.

Narrative lock-in begins when the story stops being a live explanation and becomes the frame through which everything else is interpreted. New information arrives, but it no longer has the same power to revise the picture. It is absorbed into the story the company already knows how to tell.

Sometimes this is the founding story: why the company exists, why the market needs it, why this team will be different. Just as often, it is a smaller story that forms inside a project, a product decision, or a technical assumption. The scale matters less than the function. Once the story hardens, it starts shaping what the organisation can see, what it explains away, and which questions no longer feel necessary to ask.

For that reason, narrative lock-in matters early. It distorts learning before the damage is obvious. Smart, sincere teams can keep moving while decisions, architecture, and months of work become organised around an explanation that no longer deserves the confidence being placed in it. By the time the contradiction becomes visible, the company is often no longer correcting a misunderstanding. It is paying for work shaped inside the wrong frame.

One farm-automation project made this visible to me.

The product was built for large rural properties, the kind of places where checking a gate, pump, or piece of field equipment could mean hours in a vehicle. On some properties, it could be half a day. The idea was compelling because the problem was real: give people a way to monitor and control equipment remotely instead of sending someone across the property every time something needed attention.

Early in the project, before the system had fully taken shape, we had to define what real-time meant. That sounds like a technical question, but it was really a product question hiding inside technical language.

There were options. Some architectural paths would support something closer to immediate cause and effect. Others would allow meaningful latency, minutes rather than seconds, but would be simpler, cheaper, and faster to build. The company was self-funded, so those trade-offs mattered. A latency-tolerant design was discussed, documented, and chosen for reasons that made sense at the time.

Under the original story, that decision was reasonable. If a pump was scheduled to turn on at eight o'clock and came on a few minutes later, the system was still doing something useful. A few minutes of delay did not break the idea of remote farm automation.

Then the product evolved.

Live camera feeds entered the picture. Now the interface did not only issue commands into the field. It showed the field back to the user. You could see trees moving in the wind. You could look at a gate on screen. The connection might be a little jerky, but it felt live enough to create a different expectation.

That changed the product, whether anyone named the change or not.

If you can see a gate and there is a button to open it, the product no longer feels like scheduling.

It feels like control.

That was the moment the story should have been renamed.

But it was not. The old framing held. We kept understanding the system through the earlier assumption even though the experience around it had changed. No one in the company caught the problem. No one in the technical team. No one in the business team. Myself included. We were still inside the original story.

About a year into the project, one of the owners was preparing to demonstrate the system to a prospective customer. He sat in front of the interface, looking at a live image of a gate on the property. The setup implied something simple: press the button, watch the gate move.

He pressed the button.

Nothing happened.

The video kept going. The gate stayed still. Enough time passed for the experience to stop feeling delayed and start feeling broken. Later recollection put the delay at several minutes before the gate eventually responded.

From the technical side, the explanation was clear. The system had done what it was designed to do.

That was the problem.

What surfaced in that moment was not simply a bug. It was a mismatch between the story the product now appeared to be telling and the system the company had actually built. The delay was visible, but it was not the deepest signal. The deeper signal was that the product experience had changed while the old story still governed the system.

By then, the decision was no longer just a decision. It had become history in the codebase.

At the time, nobody believed they were building the wrong thing. We believed we were finishing what we had started.

That is what made it expensive. The company did not knowingly choose the wrong product. It kept building coherently inside a story that no longer matched reality.

How It Forms

Narrative lock-in usually starts with a useful simplification. The team needs a way to describe the product, the customer, the market, the technical approach, or the problem being solved. A framing settles in: this is what we are building, this is the kind of problem we have, this is what matters for now. At that stage, the story may be reasonable. It helps people coordinate before everything is known.

Progress then gives the story weight. The same language appears in meetings, roadmaps, investor updates, product discussions, architecture decisions, and customer conversations. Each repetition makes the framing feel less like a working assumption and more like settled truth. Eventually the story stops being something the company is testing and becomes something the company is protecting.

Once that happens, explanation starts replacing exploration. Questions are answered inside the current frame rather than used to challenge it. Risks sound temporary. Contradictions look like execution problems. A concern raised today resembles one raised months earlier, and because the story has not changed, it receives the same kind of answer.

Nothing about this requires bad intent. The story hardens because it reduces complexity, protects momentum, and keeps the system feeling coherent while the work is moving. The danger is not that everyone starts believing something foolish. The danger is that the explanation stays useful long enough for people to miss the moment when reality has moved on.

Early Signals

Narrative lock-in rarely begins with dramatic struggle. It begins with small shifts in how uncertainty is handled. Updates sound the same even as circumstances change. Familiar phrases appear in presentations and status reports. Questions about assumptions are answered with explanations of the existing plan rather than investigation of the concern itself.

Repetition without revision is a useful signal. The same concern returns under slightly different wording, but the answer does not materially change. The team may be discussing the issue, but the discussion no longer seems capable of altering the frame. People are not refusing to think. They are thinking from inside the story.

Language can also start moving faster than the system. The product is described as if it has become something new, while the operating assumptions underneath belong to an older version of the work. What was once a scheduling tool is now spoken about as control. What was once a prototype is treated as a platform. What was once a narrow assumption becomes part of the way the company explains itself.

Curiosity often relocates rather than disappears. Doubts move into side conversations. Questions are raised with peers instead of in the forums where decisions are made. Messages are drafted, softened, rewritten, and sometimes never sent. Nothing appears openly suppressed. The organisation may look professional and orderly, but the range of acceptable interpretation has started to narrow.

The signal to watch is not only what people say. It is what no longer seems to change the story.

What Gets Misread

Coherence is the easiest thing to misread. Because the story sounds clear, the organisation can mistake coherence for learning. The same language keeps turning up, the same plan makes sense, and the same explanation holds the room together. That can feel like alignment. It may only mean the frame has stopped being tested.

Progress can be misread in the same way. Milestones, demos, technical decisions, customer interest, and investor language can all appear to confirm the story, even when they are only confirming that the organisation is getting better at performing inside it. A product can look more complete while the original assumption underneath it remains untested. A roadmap can look more mature while the central question has not been reopened.

Calm can also be misleading. If fewer people are challenging the story, it may look as if the doubts have been resolved. Sometimes they have simply moved elsewhere. The formal conversation becomes smoother because the harder questions no longer feel worth raising in that room.

What gets missed is the difference between a story that is still helping the company interpret reality and a story that is now filtering reality on the company's behalf. In the first case, the story changes when reality teaches the company something. In the second, reality is explained until it fits the story.

Hidden Cost

Narrative lock-in distorts learning. The company gathers information, but the information is interpreted through a frame that has become too stable. Evidence that should create revision becomes evidence that the current direction only needs more time, more execution, or more belief.

That cost does not stay abstract. Architectural decisions serve the existing narrative rather than the system's emerging reality. Product trade-offs are made to preserve the current explanation of what is being built. Customer signals are interpreted according to the story the company wants to be true. As this continues, optionality shrinks. Paths that might once have been explored feel disruptive because they challenge the explanation that is holding the work together.

By the time the contradiction becomes visible, the cost is rarely limited to one incorrect assumption. The story has influenced what was built, what was deferred, what was funded, what was explained away, and which questions stopped reaching the decision table. Narrative lock-in does not remove intelligence from the system. It makes certain questions harder to ask while the company still believes it is thinking clearly.

Recognition

The simplest test is whether the story still changes when reality teaches the company something.

If the answer is unclear, the story may no longer be live. A live story can be revised. It can absorb evidence without merely defending itself. It can help the company move while still leaving room for the company to learn.

Narrative lock-in begins when that flexibility disappears. The company keeps moving, but the story starts doing too much interpretive work. It tells people what new information means before they have properly looked at it. It keeps progress feeling coherent even when the underlying assumptions should be reopened.

This pattern sits near the start of the pattern arc because once the story filters the signals, later patterns become easier to form. Momentum feels more meaningful than it is. Earlier assumptions become harder to revisit. Confidence attaches itself to a picture of the system that may no longer be true.

Chapter 10
The Inevitability Illusion

When momentum is mistaken for validation.

The inevitability illusion begins when early progress starts carrying more certainty than it has earned.

A demo works. A prototype feels smooth. A customer sounds interested. A roadmap looks coherent. None of that is meaningless. In a startup, these moments matter because they give people something to build on. The problem begins when those signs stop being treated as evidence that something might be possible and start being treated as proof that success is on its way.

This chapter sits inside the Allusions section of the book, but the title uses illusion deliberately. An allusion is a signal that points the mind toward a larger story. An illusion is what happens when the mind mistakes that larger story for reality. The inevitability illusion is the distortion that forms when fragments of progress allude to a successful future so strongly that the organisation begins behaving as if that future has already been partly confirmed.

The distinction is important because this pattern rarely feels like fantasy from the inside. It feels like momentum. People are working, decisions are being made, and the direction may genuinely be promising. The illusion is not that nothing is real. The illusion is that the early evidence is allowed to answer a bigger question than it has earned the right to answer.

I saw this pattern form in a data-intelligence startup before there was a stable product.

The idea was expansive in a way that made intuitive sense. Satellite imagery, crop inspection, land analysis, image processing, and AI-assisted prediction would be pulled into a more connected platform. Instead of sending people to inspect fields, count trees, examine soil conditions, or look for signs of pest pressure by hand, the company imagined software that could make those conditions visible from above. A platform that could turn aerial imagery into useful operating intelligence did not sound fanciful. It sounded useful.

Early progress gave that future a visible shape. There were pitch decks, diagrams, wireframes, polished front-end graphics, and enough sample imagery and rough analysis to suggest that a much larger system was forming. On screen, the company could show something that looked like the product arriving. The future had colour, layout, workflow, and language around it.

That visibility mattered. It made the ambition feel less like a hypothesis and more like a destination.

From the outside, it looked like progress. The company was busy. The story was easy to understand. The work had enough shape to make people believe the future was arriving.

Inside the work, a few of us kept circling more basic questions.

What problem are we solving first?

For whom?

What is version one?

Those questions sound simple, but they are not small. They are the questions that turn a compelling future back into a product someone can actually use. In this case, the future kept answering louder than the product boundary did.

Inside the engineering work, the picture remained less settled.

Those questions should have narrowed the work. Instead, the larger platform story kept pulling the company back upward. The product was meant to support image ingestion, analysis, prediction, configuration, workflow, and industry-scale intelligence. But the first usable boundary remained hard to pin down. What would the first customer actually use? What data would enter the system? What was configurable? What still required an engineer? What was the repeatable product, and what was only a demonstration of the future?

While those questions remained unresolved, the technical choices grew more serious. Cloud environments were assembled. CI/CD pipelines were built. Container orchestration and heavyweight platform decisions were justified by the scale the company expected to reach. None of that was foolish in isolation. Serious infrastructure can be the right choice when the product and market have earned it. The problem was sequence. The company was building conditions for scale while the first product boundary remained unclear.

The sharpest moments came when new screens or workflow ideas were shown to the team. They looked impressive. They made the future easier to imagine. But they also raised the same practical unease underneath: what exactly were we supposed to build first?

The company had not built nothing. It had artefacts, effort, talent, code, infrastructure, demos, and a story people could believe in. The problem was interpretation. Visible movement had started carrying the emotional weight of proof. The future looked coherent enough that present uncertainty became easier to treat as a sequencing issue.

The product did not yet have to be real in the ordinary sense. It only had to feel as if it was becoming real.

How It Forms

The inevitability illusion usually forms around real movement. Something works in a narrow environment. A customer reacts positively. A technical choice makes the company feel more serious. A piece of infrastructure, design, or product language gives the future a visible shape. Each signal is limited on its own, but together they create the feeling that the path ahead is becoming obvious.

Once that feeling takes hold, possibility hardens into trajectory. The company stops asking only whether the product can become true and behaves as if the main question is when the expected future will arrive. Gaps remain, but they are reframed as sequencing problems. Missing product boundaries, weak onboarding paths, untested deployment conditions, and unresolved customer realities are all treated as things that will be solved later because the direction itself feels settled.

This is where the pattern differs from simple optimism. Optimism leaves room for uncertainty. The inevitability illusion quietly downgrades uncertainty. It does not deny every risk, but it changes how those risks are interpreted: a foundational question becomes an implementation detail. A missing proof point becomes a timing issue. A fragile demo becomes a sign of what the product will obviously become once the rest of the world catches up.

The pattern can be especially strong in technology startups because the future often does need to be imagined before it can be built. Founders, investors, engineers, and customers all have to live with some gap between what exists now and what could exist later. The danger is that the emotional pull of the future can start organising present decisions before the product has earned that level of confidence.

Early Signals

The first thing to watch is progress answering questions it has not actually tested. A working demo is used to imply readiness. A polished prototype is used to imply product maturity. A strong pitch is used to imply market clarity. A successful internal path is used to imply external repeatability. The evidence may be real, but it is being asked to carry too much meaning.

The language often shifts next. Teams speak about future capabilities as if they exist in unfinished form rather than as assumptions being tested. Roadmaps sound less like experiments and more like a route to an outcome everyone expects. The product is described by what it will become, while the present system has yet to prove what it can reliably do now.

Questions about fundamentals also soften. What is version one? Who is the first customer? What problem are we solving first? What has been proven outside the demo path? These questions may be asked, but they no longer interrupt the momentum for long. They are acknowledged, deferred, or absorbed into the larger story of progress.

You can often hear the illusion in timing language. Not yet, soon, after the next round, once the platform is ready, once the market catches up. These phrases are not automatically wrong. Startups need sequencing. But when sequencing becomes the default answer to foundational uncertainty, the company may be moving around a question it has not answered.

What Gets Misread

Progress itself is the easiest thing to misread. Activity looks like validation. The company is busy, the product is visible, the team is producing artefacts, and the story is becoming easier to tell. From a distance, that can look like proof. Up close, it may only show that the organisation is getting better at representing a future it has not yet made operational.

Polish is often misread as maturity. A product can look coherent on screen while the system underneath depends on hard-coded paths, manual work, fragile assumptions, or a narrow environment. This is becoming easier to miss in the AI era, where convincing prototypes, fluent outputs, and polished strategic language can arrive faster than the slower proof needed for real customers, real data, repeated use, support burden, edge cases, and messy operating conditions.

Seriousness can be misread as readiness. Heavy infrastructure, sophisticated architecture, senior hires, impressive partners, and ambitious tooling can all make the company feel more advanced. Sometimes those choices are justified. Sometimes they are the organisation building the conditions for scale before it has built something worth scaling.

Interest can be misread as demand. A customer who is curious, an investor who leans forward, or a market that likes the category can all make the future feel closer. But interest in the story is not the same as durable demand for the product. Attention can make a direction feel inevitable without proving that the product can survive contact with the world it claims to serve.

Hidden Cost

The inevitability illusion does not remove uncertainty. It pushes uncertainty forward in time. It becomes embedded in architecture, product shape, hiring, spending, customer promises, and emotional commitment. When the missing proof becomes unavoidable, the company is no longer testing a small assumption. It is confronting a system of decisions built around that assumption.

The pattern also changes what feels responsible. Slowing down to test the foundation can look like doubt. Narrowing the product can feel like a retreat from the larger opportunity. Asking what has actually been earned can sound negative when the room has organised itself around the future. The illusion protects momentum by making verification feel like friction.

The company can stop earning confidence from the market it claims to be serving. Instead of learning through repeatable customer use, real deployment conditions, and operational feedback, it keeps building around the success it expects. The work may be intelligent and committed, but it is increasingly shaped by a future that has not yet paid rent in reality.

This is where the pattern becomes expensive. It lets a company prepare for scale before it has built repeatability. It lets a team optimise for a future before version one is clear. It lets people believe they are closing the gap when they may only be moving the gap further down the road.

Recognition

The question is not whether the company is making progress — it is what that progress has actually proved.

Has the product earned confidence through real use, or has confidence gathered around representation? Has the first customer path become repeatable, or does each new case still require custom effort and special interpretation? Are the hardest questions being answered by evidence, or by the belief that they will make sense once the future arrives?

The inevitability illusion forms when early progress starts making success feel more settled than it is. What helps is not pessimism — it is earned confidence. Progress should increase curiosity before it increases certainty. It should sharpen the next question, not end the need to ask it.

Once a company can no longer tell the difference between what has been shown, what has been proven, and what has merely become easier to believe, the future has started governing the present. The illusion has taken hold.

Chapter 11
Borrowed Confidence

When credibility earned elsewhere is trusted here.

Borrowed confidence forms when a company starts trusting external credibility more than the present reality of its own system.

It rarely looks reckless. In most cases, it looks sensible. A respected hire joins the team. A serious vendor is attached to the work. A capable agency takes responsibility for delivery. A proven tool, platform, or methodology enters the picture. Each of those things may genuinely improve the odds. Experience matters. Outside capability matters. Prior success can be highly relevant.

The pattern begins when the reassurance starts doing more work than the evidence. The company stops asking what the current system can actually do, and starts drawing confidence from who is involved, what they have done before, or what is attached to the project. The confidence may be real. It may even have been earned somewhere else. But it has not yet been earned here.

That distinction is the centre of borrowed confidence. A company can mistake proximity to strength for strength itself. It can be close to serious people, serious tools, serious partners, serious capital, or serious language, and still not have a system that is structurally sound.

I saw this pattern in a product built for the education sector that depended on external infrastructure and external data it did not control.

The dependency was not a secret. Within the first few months, it was already clear that the incoming data was unreliable enough to threaten the product built on top of it. The issue was not a cosmetic defect. The product needed that upstream information to be dependable. Without it, the user-facing system could look plausible for a while, but it could not be trusted in the way the company needed.

A serious vendor sat behind that dependency. That mattered more than anyone admitted at the time. If the data was unstable, the assumption was that a vendor of that calibre would eventually sort it out. The problem was treated as temporary not because the product had earned that confidence, but because someone outside the company seemed substantial enough to carry it.

A phrase settled into the background of the project: the vendor will sort it out. It was rarely stated as a formal strategy, but it kept doing strategic work.

That was the borrowed confidence.

The company kept moving as though the external fix were already part of its own execution. Roadmaps continued. Product work continued. The dependency remained known, but the system was not being designed around the reality of that weakness. Very little was built to compensate for unreliable input. No serious guardrails. No real fallback logic. No meaningful intelligence for handling bad data as a normal operating condition.

Two years later, the problem still had not been resolved.

By then the product had accumulated around the dependency instead of accounting for it. The company had spent time, attention, and belief as though the infrastructure would become trustworthy before the business ran out of room. In the final stretch, there was a scramble to retrofit protections into the product under pressure, but that is a poor time to begin designing for a condition the system has known about for years.

The vendor's answer, in effect, was that another major investment might improve the situation. It might. There were no guarantees.

What had been borrowed was not only technical confidence. It was confidence in dependency. The company had started treating someone else's roadmap as if it were part of its own capability.

The quiet danger of borrowed confidence is that it can make a fragile external dependency feel like progress already owned.

How It Forms

Borrowed confidence usually appears when uncertainty is uncomfortable. The company may be struggling with technical debt, unclear product ownership, delivery pressure, weak internal capability, or a system that no longer inspires trust. Into that environment comes something that feels stronger than the current state of the work: a senior hire, a well-known consultancy, a major vendor, a respected platform, a polished agency, or an AI tool that appears to make the hard part easier.

The attraction is understandable. When a system is fragile, external credibility can feel like relief. It gives the organisation a way to believe that the problem is now in better hands. The discussion shifts from the messy reality of the current system to the capability of the people, tools, or partners now attached to it.

That shift can be useful if it leads to clearer diagnosis and better design. It becomes dangerous when it dulls the organisation's responsibility to inspect its own conditions. The company behaves as if an external fix has become part of its own execution. A vendor will stabilise the platform. A new hire will reset the architecture. A delivery partner will absorb the complexity. A tool will compress the learning curve. The bet may never be stated that plainly, but the decisions reveal it.

The pattern tends to follow naturally from the previous allusion patterns. Once a story has hardened, and once early progress has made the future feel more settled than it is, the organisation becomes more receptive to anything that appears to confirm the path. External credibility then arrives not only as help, but as reassurance that the story can continue.

Early Signals

Borrowed confidence announces itself when reassurance arrives faster than diagnosis. Concerns are raised about the present system, but the answer quickly turns toward who is involved or what has been attached to the work. They have done this before. The vendor will fix it. The agency has a process. The new lead knows how real platforms are built. The tool will make this faster.

Risk also starts being discussed differently. Instead of asking what the system is currently showing, the organisation starts asking whether the external party is credible enough to make the concern less important. Known constraints are softened because someone else now appears to own them. Fragile dependencies feel more acceptable because a serious name is associated with them.

Transfer language can carry the pattern. A person succeeded in one company, so their judgement is assumed to transfer cleanly into another. A vendor solved a problem in one environment, so its roadmap is treated as reliable in this one. A tool works impressively in a demonstration, so the organisation assumes it will hold under real operating conditions. The confidence moves across contexts faster than the proof does.

The signal to watch is not whether the outside party is competent. They may be. The signal is whether their credibility is reducing the organisation's willingness to design for the constraints it knows are real.

What Gets Misread

Capability is the usual misread. A capable person is mistaken for a correct diagnosis. A strong vendor is mistaken for a stable dependency. A polished agency conversation is mistaken for a coherent delivery system. A respected brand is mistaken for reduced risk. In each case, something real is being seen, but it is being allowed to answer the wrong question.

Prestige can be misread as accountability. In larger organisations, attaching a serious consultancy or vendor can make a project feel institutionally safer before the underlying delivery reality has changed. The reassurance is not only that capable people are involved. It is that responsibility now appears to sit somewhere more acceptable. The brand lends confidence, and sometimes even the feeling of accountability, but the project has to survive the decisions being made inside it.

Familiarity can also be misread as fit. A leader who has succeeded elsewhere may bring useful judgement, but their previous environment had its own conditions: different people, different market pressure, different technical history, different constraints, and different timing. What worked there may not map cleanly here. Borrowed confidence hides the question of transferability.

AI can create a newer version of the same misread. A tool can make a prototype, plan, or product direction look more mature than it is. It can produce fluent outputs, polished interfaces, and convincing technical movement before the underlying system is stable, trustworthy, supportable, or ready for real use. In some cases the company is not even borrowing confidence from its own intelligence, but from intelligence it has integrated and does not own. The organisation may then mistake orchestrated access to capability for capability it has actually built, understood, or earned.

Hidden Cost

Borrowed confidence can stop a company designing for the conditions it knows are true. Known fragility becomes tolerable because a vendor is expected to improve. Weak internal understanding becomes tolerable because a senior hire is expected to lead. Delivery uncertainty becomes tolerable because an agency appears to have the process under control. The current system is allowed to remain under-examined because confidence has been imported from somewhere else.

That cost compounds because borrowed confidence changes the questions people ask. Instead of asking what the present system is telling them, teams ask whether the people involved are impressive enough to justify continuing. Instead of testing the dependency, they trust the roadmap. Instead of building guardrails around known constraints, they wait for the external world to become more convenient.

The result is not only technical risk — it is decision risk. Restarts feel more credible than they have earned. Vendor promises feel safer than internal evidence supports. Outsourced work feels more controlled than the actual delivery chain may be. Tools feel like they have reduced complexity when they may only have moved that complexity somewhere less visible.

When the gap becomes clear, the company may have spent runway, trust, and attention around a confidence it did not own. The external party may still be capable. The mistake was letting their capability substitute for the company's responsibility to understand its own dependencies, constraints, and operating reality.

Recognition

Borrowed confidence is easiest to see when the company asks one direct question: what confidence have we earned here, in this system, under these conditions?

If the answer depends mainly on who is involved, what they have done before, or what brand, tool, platform, or partner is attached to the work, the company may be borrowing more confidence than it realises. The issue is not whether external help is valuable. It can be. The issue is whether that help is increasing the organisation's ability to see reality clearly, or making reality feel safer than it is.

Borrowed confidence does its deepest harm when it lets a company outsource discomfort without outsourcing responsibility. The system has to prove itself. The product has to survive its real environment. The team has to understand the dependencies it is taking on. The company has to design for the world as it is, not the world it hopes its partners, tools, or hires will create.

Distrust is not the answer. Ownership is. Use the strong hire, the vendor, the agency, the platform, or the tool, but keep asking what has been proven in the current conditions. Confidence that has been earned elsewhere can inform the work. It cannot replace the work.

Chapter 12
From Allusions to Accumulation

None of the patterns in this section announce themselves as problems.

They appear while things are still moving, while progress is still visible, while optimism still feels justified. Each one emerges from sensible responses to pressure, uncertainty, and ambition. Nothing here requires bad intent, incompetence, or recklessness.

This makes them easy to miss.

In isolation, each pattern feels benign. Even responsible. But they rarely remain isolated for long.

Narrative lock-in stabilises the story of the system and resists challenge. New information is interpreted through the existing narrative rather than examined on its own terms.

The inevitability illusion reframes uncertainty as temporary friction. Early momentum looks like proof that the current direction will eventually succeed.

Borrowed confidence draws reassurance from credibility earned somewhere else. A capable person, a serious vendor, a strong partner, a prior success, or a polished tool makes the current system feel safer than it has proved itself to be.

Together these patterns reshape perception long before they reshape outcomes. They change what evidence seems to mean.

Up to this point the changes are mostly interpretive. They affect how the organisation sees its environment and understands its progress. From the inside, nothing appears unstable. Work continues. Milestones are reached. Demonstrations succeed. Confidence remains intact.

But interpretation does not remain abstract for long.

Eventually those perceptions shape decisions. Misread evidence becomes trade-offs, deferrals, priorities, and system shape.

Small trade-offs are accepted. Shortcuts feel justified. Issues that appear manageable today are deferred until later. None of these decisions are dramatic. Each one makes sense in the moment.

Taken individually, these adjustments look like ordinary progress under pressure.

Taken together, they accumulate.

In weather terms, the pressure changes described in this section have begun forming a system. Nothing violent has happened yet, but the atmospheric conditions that make the storm possible are now in place.

Part III turns to what happens next.

Once belief hardens and uncertainty is deferred, the pressure does not disappear. It concentrates. It compounds. It shapes decisions, systems, and trade-offs beneath the surface.

Part III

The Accumulation

"The risk of accumulation is not recklessness. It is reasonableness repeated."

Chapter 13
How Strain Accumulates

Accumulation is rarely dramatic.

It unfolds in the long middle stretch, where nothing is obviously wrong, yet constraints quietly compound beneath the surface. Decisions stack. Shortcuts normalise. Trade-offs settle into habit. What once felt like pragmatic flexibility slowly hardens into structure.

This is the phase where organisations are often most confident in their direction.

Progress is visible. Delivery continues. Metrics move. From the outside, the system appears to be working. From the inside, there is little incentive to slow down.

In startups, this period reads as momentum. Movement substitutes for clarity. Continuity stands in for coherence. Early wins justify deferred fixes, and deferred fixes gradually lose their provisional status.

Alongside technical debt, something less visible accumulates: fatigue, friction, role blur, and the quiet loss of shared authority. Processes remain visible while the real work starts moving around them. Incentives drift away from the values that once anchored them. Responsibility stretches across growing systems, diffusing ownership while increasing dependency.

As teams scale, decisions once made close to the work are mediated through layers, abstractions, and handoffs. Context thins. Assumptions travel further than evidence. What was once shared understanding becomes loosely aligned interpretation.

The risk of accumulation is not recklessness.

It is reasonableness repeated.

Pressure rewards continuity. Optimism dampens hesitation. The system learns which questions slow progress, which compromises preserve momentum, which processes can be bypassed, and which kinds of work attract attention.

Accumulation reshapes baselines over time. What began as an exception becomes precedent. What was temporary becomes assumed. Practices persist not because they are examined and affirmed, but because they no longer attract attention.

What forms instead is a system shaped by yesterday's compromises, optimised around assumptions no longer revisited.

When strain becomes visible, the conditions that produced it may have been in place for years.

Part III examines how those reasonable decisions accumulate, not as isolated mistakes, but as patterns that quietly reshape the systems teams depend on. Incremental compromise shows how temporary trade-offs become design. Process erosion shows how the agreed way of working loses authority. Incentive drift shows how the reward signal pulls effort away from structural health.

Chapter 14
Incremental Compromise

When temporary trade-offs become design.

Incremental compromise begins when temporary, practical trade-offs keep being made under pressure until they stop being temporary and start becoming design.

It rarely looks like someone deciding to lower the standard. More commonly, it looks like one more reasonable adjustment in a system that needs to keep moving. A fix will do for now. A workaround will hold. A proper pass can happen later, after funding, after the demo, after launch, after the next push.

Each step can make sense on its own. Time matters. Money matters. Momentum matters. In a startup, refusing every compromise can be just as dangerous as accepting too many. The pattern is not compromise itself. The pattern is what happens when the organisation stops distinguishing between a temporary trade-off and the way the system now works.

Incremental compromise belongs at the start of Accumulation because the early allusion patterns distort what the organisation thinks it is seeing. This is where distorted interpretation gathers material form. Belief, pressure, and reassurance become architecture, process, product behaviour, and operating habit.

I saw this pattern in one fitness-app company through a product that always seemed close.

The company had a clear idea and a market it understood. The app connected to wearables and smart shoes, showed performance charts, and carried an AI layer that sounded like a genuine differentiator. In screenshots and controlled demonstrations, it could look like a real product taking shape. The next milestone always seemed within reach: a credible MVP, a funding conversation, a version that would prove the company was ready for the next stage.

Underneath, the system was more fragile than the surface suggested.

Data did not always display correctly. Buttons failed. Integrations were unreliable. The AI capability was inconsistent and slow. Some flows could be made to work if the path was narrow enough, but the product did not yet behave like something people could trust without careful handling.

Compromise becomes seductive in that gap. It did not look infinite. It looked bridgeable. Another fix might do it. Another patch might make the demo smoother. Another push might get the product close enough to justify the next conversation.

That was part of the trap. The MVP was not just a delivery milestone. It was supposed to unlock the next stage: funding, legitimacy, and a future the company could still believe in. Every push therefore carried more strategic weight than it should have.

So the company kept trying to get it over the line.

One bug would be fixed and another would appear. A flow would hold together long enough to restore confidence, then fail somewhere else. The product did not collapse all at once. It kept preserving the appearance of closeness while the work needed to make it trustworthy remained deferred.

Eventually, they brought in a senior engineer to make sense of the system. He introduced more disciplined bug tracking, worked through the product systematically, and surfaced hundreds of issues with evidence and explanation. From an engineering point of view, that was progress. It made the deferred work visible. To the founders, it landed differently. It looked like proof that the product was broken.

The distinction matters. When visibility arrives inside a system shaped by accumulated compromise, it does not always feel like help. It can feel like indictment. The audit was not only showing defects. It was showing the cost of many earlier decisions that had preserved the appearance of closeness without creating the conditions needed for confidence.

The company started again.

On the surface, that sounded decisive. But a restart does not automatically remove the pattern that made the first version hard to trust. Ownership did not become clearer. Project memory did not become stronger. The urgency remained. Some old fragments and assumptions found their way into the new system in altered form.

So the product became close again: close enough to keep hope alive, but not stable enough to carry the promise cleanly.

By the time I was asked to look at it, the handover itself had become part of the evidence. Documents, screenshots, old reports, repository links, partial explanations, and outdated advice were all pushed together without a coherent account of what was current, what was obsolete, and what actually worked.

Each decision in that sequence had pressure around it. Money mattered. Time mattered. The next push might genuinely have been enough. The pattern was hard to stop because each compromise preserved the possibility of success for a little longer, while also making eventual recovery harder, slower, and more expensive.

How It Forms

Incremental compromise usually forms in conditions where the company is trying to preserve motion. There is a deadline, a customer conversation, an investor update, a demo, a launch, or a team under pressure to prove that progress is still happening. In that environment, the immediate question is rarely, "What kind of system are we becoming?" The immediate question is, "How do we get this over the line?"

The first compromise can feel controlled. The team knows it is not ideal. They name it honestly enough. They might even agree to revisit it later. At that point, the trade-off is visible. It is a conscious decision made under constraint.

The pattern begins when the same kind of decision keeps returning. Documentation can wait. Tests can wait. A cleaner boundary can wait. A hard product question can wait. A brittle integration can be managed manually for now. The company does not feel as if it is drifting. It feels as if it is sequencing the work sensibly.

Later becomes a warehouse for everything the company does not yet have the time, clarity, or courage to confront. Each item placed there makes the next one easier to place. What began as exception becomes precedent. What began as a patch becomes a dependency. What began as a temporary path becomes the path other work is built around.

Early Signals

Incremental compromise appears in language before it appears in the architecture. Phrases like "not yet," "for now," "we'll come back to it," and "let's just get this over the line" do not sound careless. They sound practical. They are worth listening to for that reason.

The signal is not the phrase itself. Startups need temporary choices. The signal is repetition. If the same deferral keeps appearing in different forms, or if more parts of the system depend on promises to return later, the organisation may no longer be managing temporary compromise. It may be normalising it.

Localised caution is another clue. Fixes become smaller, narrower, and more defensive because no one wants to disturb an area they do not fully understand anymore. Certain parts of the system develop reputations. Everyone knows they are fragile. Everyone works around them. The knowledge becomes cultural before it becomes formal.

The pattern is also visible when visibility feels threatening. A proper audit, bug list, architectural map, or product review should help the company see the way forward. Inside a system built from accumulated compromise, that same visibility can feel like indictment.

The organisation wanted progress. What it receives is evidence of how much has been deferred.

What Gets Misread

Continuity is easy to mistake for control. Because the product moves, the company assumes the compromises are under control. Demos may work. Customers may be interested. The team may keep shipping. From the outside, the organisation appears to be preserving momentum. Underneath, it may be spending more and more energy preserving the possibility of movement.

Speed can be misread as progress. A shortcut can make the next visible step arrive sooner, but it may also make later steps harder, slower, or more expensive. The company may not really be gaining speed. It may be borrowing speed from the future and leaving repayment for a version of the organisation with less time, less trust, and less margin.

Practicality can also be misread. A decision can be practical in the moment and damaging as a repeated pattern. This is what makes incremental compromise hard to challenge. Each decision can be defended by the pressure in front of it. The harm only becomes obvious when the sequence is viewed as a system.

Even closeness can be misread. A product that is always nearly ready can keep justifying another push. The MVP is close. The demo is close. The next fix might be enough. After enough rounds, closeness stops being evidence of progress and starts becoming part of the trap.

Hidden Cost

Incremental compromise later shows up as a gradual loss of ease. The system may function, but everything takes more effort. Changes require more context. Fixes require more caution. New people need more informal explanation. More work depends on memory, patience, and local knowledge.

This cost accumulates quietly because the company can point to visible progress. The problem is not immediate collapse; it is that each compromise reduces future manoeuvrability. The organisation becomes less able to change direction, less able to onboard customers cleanly, less able to explain its own system, and less able to recover without disturbing something else.

Eventually, the company begins spending more energy preserving motion than improving structure. A feature is not just a feature; it is a negotiation with old shortcuts. A fix is not just a fix; it is an attempt not to break whatever has grown around the compromise. The product becomes harder to trust, harder to change, and harder to recover.

The deeper cost is that temporary stops being temporary. It becomes design, not because anyone chose it as design, but because enough later decisions depended on it being there. By the time the organisation decides to repair it, it is no longer fixing one shortcut. It is unwinding a chain of reasonable decisions that have become part of the product's shape.

Recognition

Incremental compromise is best tested by asking what is still genuinely temporary, and what has quietly become the standard.

If the answer is uncomfortable, that is the point. Incremental compromise does not ask whether a team has ever made a trade-off. Every serious team has. It asks whether the organisation still knows which trade-offs it is carrying, why they were made, and what it would cost to reverse them.

Perfection is not the point. A startup that refuses all compromise will struggle to move. The healthier discipline is to keep temporary decisions visible enough that they do not become invisible architecture. Name the compromise. Name the reason. Name the condition under which it must be revisited. If those conditions never arrive, be honest that "later" has become the plan.

Incremental compromise causes the most lasting harm when each reasonable choice makes the next one easier, until the system is shaped less by deliberate design than by what was easiest to defer under pressure. Accumulation begins this way: not with one reckless decision, but with many practical ones that were never brought back into view.

Chapter 15
Process Erosion

When the process remains visible but stops carrying authority.

Process erosion begins when the agreed way of working remains visible, but stops carrying authority.

The process is not abolished. The tickets may exist. The stand-ups may happen. Reviews, signoffs, definitions of done, and delivery rituals may all remain in place. From a distance, the organisation can look disciplined.

What changes is the practical rule underneath. The real workflow becomes the one shaped by urgency, proximity to power, and whatever gets the visible result over the line fastest.

This separates process erosion from incremental compromise. Incremental compromise is about the trade-offs that accumulate inside the system. Process erosion is about the coordination layer that was supposed to make those trade-offs visible, deliberate, and recoverable. When that layer loses authority, the organisation does not only carry more compromise. It loses the shared mechanism for seeing what it is carrying.

I saw this happen in one startup.

I was part of an engineering team trying to establish basic development discipline while the company pushed toward investor conversations, pitch material, and demos. The process was not heavy. It was the normal scaffolding a small technical team needs if work is going to stay visible: shared priorities, review paths, some agreement about what was being changed, and enough coordination that the rest of the team could understand what had moved.

Then visible progress became urgent, and instead of moving the work through the agreed process, the founder went directly to a junior developer for weekend and after-hours demo work. Screens changed.

Demo features moved. The product looked more alive by Monday morning.

But the team had not moved together.

When the rest of us arrived back at work, we found that decisions had effectively been made outside the process we were trying to build. The issue was not only that code had changed. It was that the organisation had just been taught something about authority. When the work was visible enough, urgent enough, or close enough to the founder's attention, the process could be bypassed.

Nothing about the official process changed on paper. That was the point. The process existed. It just mattered less than the path around it.

A later version of the same lesson made the rule even clearer. A more disciplined release cadence was discussed: define the scope, hold the line, and let unfinished work fall into the next release instead of dragging everything forward under pressure.

That sounded reasonable. It was even accepted in principle.

But only with an exception.

If something was needed for a presentation, a client conversation, or a change in direction, the process still had to be breakable. The cadence could exist, but not where it constrained the visible moment that mattered most.

That is how process erosion becomes cultural. The organisation does not say process is unimportant. It says process matters, then teaches everyone where it will not hold.

How It Forms

Process erosion usually begins with exceptions that seem reasonable. A review is skipped because the work is only for a demo. A founder goes directly to the person who can move fastest. A definition of done softens because the team needs something visible by Monday. A bug is left in a code comment because writing it up properly feels slower than telling the person beside you.

These moments can all be defended by the pressure around them. In a young company, process has to be light enough to preserve movement. A team that turns every decision into ceremony will slow itself into irrelevance. The problem is not that exceptions occur. The problem is what the organisation learns from them.

If an exception produces the visible result, it becomes easier to repeat. If bypassing review gets praised, review becomes optional. If a side channel solves the urgent problem, the side channel looks more useful than the formal one. If the person with power keeps overriding the process, the process may be described as important, but everyone understands when it does not apply.

The formal system and the lived system drift apart over time. The official process becomes the map. The real process becomes the path people actually take to survive pressure. That gap is where erosion lives.

Early Signals

Process erosion does not usually begin as obvious disorder. It begins as a subtle split between what the company says the process is and how work really moves.

People update tickets, but the important decisions happen in calls, messages, hallway conversations, or direct founder requests. Standups happen, but the team is reporting work redirected elsewhere. Reviews exist, but they happen after the practical decision has been made. Signoff exists, but it no longer constrains what gets built.

Exception language is a strong clue. "Just this once." "Only for the demo." "We can clean it up later." "Don't block the team." "It's already basically done." These phrases often sound collaborative. They can also teach the organisation that discipline is something to respect when pressure is low and bypass when pressure is high.

Where people place their attention tells the same story. If the team spends more time explaining what happened than working from a shared record of what was agreed, the process is no longer doing its coordinating job. If new people need informal guides to understand how work really gets done, the process may exist, but the knowledge required to use it has moved into people.

A software team can see this in very small places. On one large project, formal bug and risk systems still existed, but some of the real coordination burden had drifted closer to the code: comments, local scripts, verbal handovers, and private explanations about which shortcuts were safe. Those fragments were practical in the moment. They were also a sign that the official process no longer carried the full truth of the work.

The most important signal is authority. When process has authority, it can slow a decision, expose a risk, or force a question back into view. When process has lost authority, it becomes a record of movement rather than a constraint on movement.

What Gets Misread

Process erosion can look like flexibility. The organisation tells itself it is staying lean, avoiding bureaucracy, or empowering people to make decisions. Sometimes that is true. Early companies should not copy the weight of large-company process before they have earned the need for it.

But flexibility and erosion are not the same thing. Flexibility adapts the process while preserving shared understanding. Erosion bypasses the process while preserving the appearance that shared understanding exists.

Output creates the next misread. If work keeps shipping, leaders may assume the process is healthy enough. The product moves. Demos happen. Customers see progress. From altitude, the system appears responsive. But output can continue long after the coordination layer has started to fail. People can keep a fragile process alive through memory, effort, favours, and local knowledge.

That effort is easy to miss because it often looks like commitment. A senior person explains the same context repeatedly. A developer remembers which shortcut is safe. A project lead translates between what the ticket says and what the team knows it really means. None of that appears as process failure. It appears as people doing whatever it takes.

Documentation can mislead in a similar way. A company may assume that because artefacts exist, process exists. Tickets, user stories, bug lists, status updates, and handover notes are useful only if they govern shared action. If they become paperwork people maintain after decisions are made elsewhere, they stop being process and become theatre.

AI can accelerate this misread. Fast output can make a thin process feel adequate because there is always something visible to inspect. But visible output is not the same as shared reasoning, ownership, validation, or recoverability. If AI helps a team move faster while weakening the moments where people clarify, review, and own decisions, it does not fix process erosion. It speeds it up.

Hidden Cost

Process erosion leaves the organisation less reliable than it appears. Once process loses authority, coordination depends on who remembers what changed, who knows the safe path, who was present for the conversation, and who has enough local context to fill in the gaps. The company may have tools, rituals, and language of discipline, but the real operating system has moved into individuals and side channels.

That creates fragility. Work becomes harder to hand over. Decisions become harder to reconstruct. New people take longer to become useful because the written process does not explain how the company actually works. Risk becomes harder to escalate because the places where risk should appear no longer carry the full truth of the work.

The cost also lands emotionally. When the process no longer protects the team, people learn that discipline is negotiable but consequences are not. They remain responsible for outcomes, but the mechanism that should make work legible has been weakened. That creates frustration, quiet resignation, and eventually the sense that the only way to survive is to work around the system as well.

At scale, erosion creates a dangerous illusion for leaders. The formal view remains orderly while the lived work becomes increasingly improvised. Reports, boards, and status updates can look cleaner than reality because the mess has moved somewhere else. When the gap becomes obvious, the company may discover it does not have one process. It has several partial processes, each held together by different people.

Recognition

To recognise process erosion, look at where work really moves when pressure rises.

If the answer is different from the official process, the next question is whether that difference is deliberate and visible, or informal and hidden. Healthy teams adjust process openly. Unhealthy systems let the official process remain in place while the real process goes underground.

The answer is not heavier process for its own sake. The discipline is to preserve enough authority in the process that important work cannot quietly bypass visibility, review, ownership, or shared understanding. A process can be light and still have authority. It can be simple and still tell the truth.

A useful test is whether the process can still interrupt momentum. Can it surface an unresolved question before the demo? Can it make a skipped review visible? Can it force a decision back to the people who own the consequences? Can it tell a new person how the work actually moves, without requiring a private translation layer?

Process erosion does its deepest damage when the company keeps the language of discipline while teaching people that discipline applies only until it slows visible progress. Once that lesson takes hold, the process has not disappeared. It has become decorative. The real system is being run somewhere else.

Chapter 16
Incentive Drift

When the work follows what gets rewarded, not what most needs doing.

Incentive drift begins when the organisation's energy starts following what is easiest to admire, report, sell, fund, or defend as progress.

The incentive is not always written down. Sometimes it lives in a dashboard. Sometimes it lives in founder attention. Sometimes it lives in investor questions, sales conversations, board pressure, internal status, market excitement, or the simple fact that some work makes the room feel more alive than other work does.

The pattern is hard to catch because the company may be working hard. It may believe in the mission. It may be making decisions that sound rational in isolation. But the centre of gravity moves. Work that produces visible proof attracts more energy than work that protects structural health.

Incentive drift is not the same as process erosion. Process erosion is about the agreed way of working losing authority. Incentive drift is about the reward signal changing what people choose to work on, celebrate, tolerate, or ignore. When the two combine, the company does not just bypass discipline. It learns to bypass discipline in the direction the incentives are pulling.

I first saw this pattern clearly in an early golf-booking startup.

The company was building software for golf courses to manage tee times, bookings, and related operations through a shared platform. In the office, we could make the product look smooth and convincing. Sales demonstrations ran on carefully prepared laptops. The full environment could be carried locally, with the interface, database, and supporting pieces all close enough together that the product felt fast. A booking could be entered on one side of the system and appear almost immediately on the other. Screens responded. Edits looked clean. The future felt visible.

The real deployment environment was less forgiving.

Many golf courses were not sitting inside city-office conditions. Some relied on slow or unreliable connections, including satellite links. In those environments, the same product could become sluggish enough to frustrate the people using it. A golf pro trying to change a booking at the counter could click, wait, make the change, save, and then wait again while the system caught up. The product existed. It could be demonstrated. But the operational reality was much less impressive than the version that travelled well in sales conversations.

The engineering problem was not mysterious. Those of us close to the work understood that the system needed to become less dependent on fast, reliable network conditions. There were discussions about caching, reducing round trips, and changing parts of the architecture so the product could tolerate the world it actually had to operate in.

That work was important.

It was also difficult, mostly invisible, and hard to sell.

The harder work was slower to explain and almost impossible to sell. The more glamorous work could be shown in a conversation, a slide, or a demo and immediately feel like momentum.

Meanwhile, attention kept moving toward work that made the product sound more advanced. Late in the company's life, energy flowed into a voice-recognition booking feature. Golfers would be able to call a number, speak to an automated system, and make a booking without talking to the clubhouse. For the period, that sounded genuinely futuristic. It gave sales conversations more spark. It made the company feel as if it was building ahead of the market.

The problem was not that voice recognition was silly. It was plausible. It was interesting. In another context, it might have been exactly the kind of feature that opened a useful future.

The problem was the order of attention.

The product still struggled in the conditions where it needed to be used, but the work that would have made it more dependable did not create the same visible lift. Reliability did not photograph well. Fewer network calls did not make the room feel more excited. Local caching did not sound like the future in the way voice recognition did.

So we kept building the future while the present remained underbuilt.

The incentive was not written down anywhere.

It lived in what drew admiration.

How It Forms

Incentive drift usually forms in the gap between stated priorities and rewarded behaviour.

A startup might say reliability matters, but celebrate the feature that looks impressive in a demo. It might say customer readiness matters, but reward the milestone that turns a chart green. It might say focus matters, but give the most attention to the newest market opportunity. It might say product truth matters, but become most animated by whatever makes the company feel current, ambitious, or investable.

At first, that pull is understandable. Startups need proof. They need momentum. They need something visible enough to persuade customers, investors, partners, and themselves that the effort is becoming real. Visible progress matters. The drift begins when visible progress becomes easier to reward than the quieter work that makes progress durable.

People learn the practical incentives over time. They learn which work gets attention, which questions create friction, which risks slow the story, and which achievements make leaders feel confident. No one has to announce the rule. The organisation teaches it through response.

Incentive drift becomes structural when the team does not choose the wrong task once. It begins optimising for the signals the environment keeps rewarding.

Early Signals

Incentive drift often appears when necessary work keeps needing to justify itself while visible work attracts automatic energy.

Reliability work needs a business case. Testing needs defending. Documentation needs to wait. Customer onboarding issues need more evidence. Meanwhile, a feature that demos well, a metric that looks good, a partnership that sounds strategic, or a market angle that feels exciting moves faster because it creates immediate confidence.

Status can begin replacing substance. The company starts managing the colour of the report, the completion of the milestone, the appearance of adoption, or the story attached to the work more carefully than the underlying outcome. The project is green, but the user experience is weak. The metric improves, but the behaviour behind it is unclear. The tool is adopted, but no one asks whether it is making the product better.

Repeated phrases can reveal incentive drift: "This will look good for investors." "We need something to show." "The board wants to see movement." "Customers will love this." "Everyone is talking about this space." None of those sentences is automatically wrong. The signal is whether they keep outranking the less glamorous questions: does this solve the right problem, does the system support it, can customers actually use it, and what are we neglecting to make this visible thing happen?

A modern version appears when the metric itself becomes the thing people serve. Ticket count, deployment frequency, platform adoption, or some other measurable signal can become a proxy for progress. Once that happens, people learn to optimise for the proxy rather than the outcome it was meant to represent.

What Gets Misread

Incentive drift can look like ambition. The organisation feels energetic. People are chasing opportunities. The product is expanding. The company is responding to the market. From the outside, it can look like momentum.

Sometimes it is momentum. The question is whether the movement is strengthening the product or merely strengthening the story that the product is becoming important.

Responsiveness creates a subtler misread. A startup needs to respond to signals from customers, investors, and the market. But responsiveness becomes drift when every external signal is allowed to redirect internal attention before the company has clarified what it is actually learning. A question from an investor becomes a roadmap item. A fashionable technology becomes a strategy. A sales objection becomes a rushed feature. The company looks adaptive, but its centre of gravity is being pulled around by whatever carries the strongest immediate reward.

Morale can be misread too. Enthusiasm can feel like evidence. A smart room, a lively demo, a new opportunity, or a shared sense that the company is onto something can all create the feeling of proof. But energy is not validation. Admiration is not adoption. Excitement is not product-market fit. If the organisation treats the emotional lift as evidence that the direction is right, incentive drift has begun to shape interpretation.

There is also a subtler misread around internal platforms, tools, and teams. A company may push work through a particular system not because the product needs it, but because the platform, budget, or function needs to justify itself. In those moments, architecture starts serving organisational incentives rather than product need.

Hidden Cost

Incentive drift teaches the organisation to produce distorted signals. It becomes increasingly effective at producing what it rewards, even when those signals are no longer reliable evidence of health. Teams get better at making work look complete, advanced, fundable, strategic, or innovative. They also get better at deferring the work that does not create the same visible lift.

That distortion compounds. Reliability work falls behind because it is harder to admire. Customer readiness slips because it is less exciting than the next feature. Product focus weakens because every attractive opportunity feels like proof of possibility. The company may move quickly, but more of that movement is now shaped by what can be defended as progress rather than by what would make the system stronger.

The cost is not only technical. Incentive drift changes what people are willing to say. If the reward sits on optimism, people become careful about surfacing drag. If the reward sits on green status, problems are softened until the report can stay green. If the reward sits on visible innovation, unglamorous work feels like a career risk. Over time, the company does not just misallocate effort. It miseducates its own attention.

Eventually, the stated mission and the lived incentives no longer match. The company still says it is building for customers, quality, learning, or scale. Its behaviour says it is building for admiration, reporting comfort, valuation, internal politics, or whatever currently looks easiest to call success.

Recognition

Incentive drift shows up in the work that gets energy before it has to argue for itself.

That line of inquiry usually reveals more than the strategy document. Look at what moves quickly. Look at what leaders ask about first. Look at which risks require repeated explanation. Look at which work gets praised, protected, funded, or forgiven. The real incentives are often visible in the organisation's reflexes.

The companion question is less comfortable: what necessary work keeps losing because it is harder to admire?

If the same quiet work keeps being postponed, the issue may not be prioritisation. It may be that the organisation has built a reward system in which that work cannot win. Reliability, testing, onboarding, architectural clarity, customer support, and operational readiness often lose this way. They matter deeply, but they do not always create the immediate emotional or political return that more visible work creates.

The aim is not to remove ambition. A startup without ambition will not survive. The discipline is to keep checking whether the reward signal still points toward the outcome the company claims to want. If a metric is rewarded, ask what behaviour it will create. If a feature is celebrated, ask what it displaces. If a platform is mandated, ask whose problem it solves. If a market opportunity starts pulling attention, ask whether it strengthens the core or gives the company a more exciting way to avoid it.

Incentive drift is hardest to reverse when the organisation keeps pursuing its mission in language while pursuing something else in practice. The work follows the reward signal. If that signal drifts, the company drifts with it.

Chapter 17
From Accumulation to Human Cost

Late at night the office lights are still on.

Most of the building is empty, but a small group of engineers remain at their desks. A deployment that should have been simple has stretched into hours of investigation, restarts, and quiet workarounds.

No one is surprised.

The system still works, but it has become delicate. Certain services must be restarted in the right order. Certain changes must be deployed carefully. Certain problems are understood but postponed because there has never quite been enough time to solve them properly.

Everyone in the room knows how the system behaves now. They have adapted to it.

From the outside the company still looks like it is moving forward. Features are released. Demos succeed. Roadmaps continue to fill with new plans.

Inside the room, however, the weight of keeping the system moving is carried quietly by the people maintaining it.

The architecture does not feel the strain. The process does not feel the strain. People do.

Accumulation rarely announces itself as damage. It settles in gradually, while systems continue to function and work continues to move. Nothing collapses outright. Decisions remain rational. Systems remain operational. People remain capable. Each adjustment makes sense in isolation, and together they create something that feels stable enough to continue.

Accumulation is difficult to interrupt for precisely that reason.

Incremental compromise turns temporary trade-offs into design. Process erosion shifts authority from the agreed way of working into memory, side channels, and local habit. Incentive drift teaches effort to follow the reward signal, even when the reward signal no longer points toward structural health. Visible momentum can continue even when many people can feel the strain building.

None of this looks like a breaking point. It looks like progress under constraint.

Momentum remains visible even as the structure carrying it becomes thinner. From the inside, accumulation feels like work being done by sensible people doing their best. From the outside, it often looks like forward motion.

But accumulation makes its deepest mark not in architecture diagrams, process documents, or status updates, but in people.

The cost of this phase is rarely paid immediately. Like pressure building inside a weather system, the strain accumulates quietly before it becomes visible. It is deferred, distributed, and absorbed. It settles into habits, expectations, and unspoken limits. It waits.

Once accumulation becomes visible, it is no longer only a question of what went wrong.

The more important question is who has been carrying what the system could no longer hold.

Because no system accumulates in isolation. The weight created here does not disappear. It moves. It lands somewhere. And more often than not, it lands with the people closest to the work.

The book turns there next.

Part IV is about that landing.

Part IV

The Human Cost

"Accumulation does not make its deepest mark in architecture diagrams, process documents, or status updates. It makes its deepest mark in people."

Chapter 18
Where the Cost Lands

The human cost emerges when pressure moves through a system and settles somewhere else.

It rarely appears where decisions are made, plans are drafted, or architecture is discussed. Instead, it is absorbed by people.

This absorption remains quiet. It does not announce itself as harm or crisis. It shows up as adaptation — narrowed focus, emotional distance, relief when someone else breaks first. The work continues, outwardly intact, while something human adjusts beneath the surface to make that continuation possible.

This is where the human cost accumulates.

I have seen it appear in a simple image: people still at their desks late in the evening, then seemingly back in the same chairs the next morning, in the same positions, as if the system had started drawing continuity directly from their bodies. Nothing had exploded. No one had announced a crisis. The work was simply continuing because people had become the buffer.

The human cost may first become visible this way. Not as collapse, but as compensation.

Up to this point, the book has examined decisions, incentives, momentum, and systems. These are discussable because they feel rational, measurable, and controllable. Beneath every deferred fix, every shortcut, every quietly lowered standard, a human price is paid.

This part examines that cost.

Not dramatic collapse stories retold later, but the erosion that occurs while everything still appears to be working. Fatigue becomes normal. Personal stretch looks like ordinary capacity. Emotional distance reads as professionalism. People protect themselves by filtering what they bring into the room long before the business shows visible signs of strain.

The chapters that follow describe three different movements inside that human cost. First, unresolved burden gets transferred into people as attention, patience, memory, judgement, care, and endurance. Then the stretch required to carry that burden stops feeling exceptional and starts becoming baseline. Finally, some of the people carrying the strain begin filtering how much signal, honesty, and expertise they bring into the room.

Teams carry this cost in longer days, private concern, repeated workaround thinking, and the gradual normalisation of strain.

In the beginning, enthusiasm, culture, and shared belief can make a company feel as though it can absorb almost anything. For a while, that energy does carry people. But as reality-load increases, energy stops functioning as margin and starts becoming something the organisation spends.

What once felt like belief starts turning into burden.

Founders carry it too, even if that reality can be harder to speak about cleanly. They carry the weight of belief, the pressure of expectations, the emotional load of keeping confidence alive, and the growing tension between what they need the company to be and what they may suspect it is becoming.

That cost can be difficult for founders to name cleanly because part of their role is to keep belief functioning even when certainty is thinning. They are expected to absorb ambiguity, convert doubt into direction, and carry a level of confidence that the wider organisation can continue building around. In healthy periods, that can look like leadership. Under prolonged strain, it can become a quieter form of isolation.

A founder may be the one most responsible for the trajectory and also one of the people most trapped inside it. The same person who keeps momentum alive may also be the least able to publicly register how much is being carried through intuition, hope, performance, or private fear. That does not excuse bad decisions. But it does help explain why some trajectories persist longer than they should. The burden is not only strategic — it is emotional, reputational, and profoundly difficult to put into ordinary language while the company is still trying to move.

You can sometimes see it before anyone names it. A founder arrives in the room short on sleep, still trying to project conviction, but thinner around the edges. The explanation becomes less steady. The reactions become sharper. The person is still leading, still carrying the story, still trying to keep the company moving, but the strain is no longer only inside the business. It is moving through the person holding the business together.

Some of that burden also follows people outside the formal system. It shows up later in sleep, mood, patience, relationships, and the private effort required to stay steady while the public story still asks them to perform confidence.

The human cost rarely appears on a balance sheet. It surfaces as repeated quiet compensation, disappearing recovery, careful meetings, softened truth, and a widening gap between what is said publicly and what is known privately. It accumulates as stress, cynicism, burnout, and quiet resentment, carried by those most invested in the outcome.

The human cost is not only a side effect of the pattern arc.

It also helps drive the arc forward.

People do not only defend brittle trajectories because the reasoning seems sound on paper. They also defend them because stepping outside the trajectory may mean confronting sunk cost, founder identity, team fatigue, reputational fear, and the possibility that months or years of sacrifice have been stabilising something far weaker than everyone hoped.

Pressure rarely disappears inside organisations.

It travels.

It moves through layers of responsibility, incentives, and distance from the work until it finds somewhere it can be absorbed. Usually, it settles with the people closest to the system — those who see the strain first and have the least authority to change its direction.

Here, the focus shifts from strategy to impact.

How pressure travels through teams. How recovery disappears. How fear reshapes behaviour. How responsibility, when paired with distance from the work, amplifies harm rather than accountability. How unresolved human debt compounds just as surely as technical debt, leaving effects that linger long after code is rewritten or companies disappear.

If the earlier parts of this book help make sense of how struggle forms, this part asks a harder question: who carries it, and at what cost over time?

The patterns described in this book do not happen above people.

They happen through them.

No venture struggles in isolation. No success is clean if it extracts more from people than it ever returns.

Chapter 19
Human Debt Accumulation

When unresolved strain gets stored in people.

Human debt accumulation begins when strain the organisation has not properly resolved starts getting stored in people instead.

The system appears to function. Work moves. Delivery may look adequate from a distance. The board, the founder, or the customer may see progress. But some of the burden has already shifted.

People become memory where process is thin. They become support where capability is mismatched. They become interpretation where requirements are vague. They become the quiet buffer between what the system demands and what it can visibly deliver.

At first this can look like commitment. Someone helps again. Someone remembers the fragile path through the system. Someone absorbs a difficult handover, explains the same thing one more time, or carries the emotional weight of keeping the room moving.

That commitment becomes stored cost over time.

This is where the arc turns toward the human cost. The earlier Accumulation patterns showed how compromise, eroded process, and drifting incentives reshape the system. Human debt accumulation shows where the unresolved load goes first. It does not disappear into the architecture. It moves into attention, patience, memory, judgement, and endurance.

One complex project made this visible through a staffing decision that looked sensible on paper, but changed the daily shape of the room around the senior people already carrying the work.

The work was already difficult. The system was moving quickly, the technical surface area was wide, and much of the project depended on senior judgement that was hard to document cleanly. Adding junior developers looked like added capacity. More hands. More throughput. A way to spread the work.

What it spread most reliably, however, was interruption.

In practice, at least one of those hires created a different kind of load.

The issue was not only code quality. Ordinary parts of the job needed repeated support: tooling, command-line work, project setup, local environment issues, and the small day-to-day fluency that lets a developer move without constantly pulling someone else out of flow. None of those requests looked unreasonable in isolation. A senior person helped. Then another. Then another.

But the help was not free.

It came out of the attention of the people already carrying the hardest parts of the system. Their own work did not shrink. Their delivery expectations did not change. They remained responsible for difficult architecture, fragile code, deadlines, decisions, and coordination. Now they were also carrying the support load required to make the extra headcount function.

That gave the project a particular feel. The visible work was still features, fixes, deadlines, and delivery. The hidden work was repeated explanation, unblocking, setup help, reassurance, and context translation, all landing on the same few people often enough to break the shape of their own day.

At one point the hidden cost was named directly: this person might be the most expensive developer in the building once the support burden was counted. Not because of salary, but because of the senior time being consumed around them.

Organisations miss this part. A staffing decision can look cheap while quietly spending the most expensive capacity the team has.

The burden also has a way of following the people who know the system best. Time away from the project was not always fully protected either. Something broke, and the person closest to the work was pulled back in. More than interruption, it is dependency made visible.

The organisation discovers that continuity is not really stored in the process, the codebase, or the staffing plan. It is stored in a person.

When the issue was raised, the hidden cost was visible to the people close to the work.

It did not materially change the staffing choice.

So the burden stayed where these burdens usually stay: in the interruptions, patience, judgement, and lost capacity of the people around the gap. Later, a broader hiring push reproduced the same pattern at larger scale. More people appeared to mean more capacity. In practice, it also meant more onboarding, more explanation, more coordination, and more senior attention being consumed.

The organisation had not only added people.

It had moved cost into people.

How It Forms

Human debt forms when an organisation solves a structural problem through human absorption rather than structural repair.

A process is unclear, so a senior person keeps translating it. A product is fragile, so one developer becomes the person who knows which changes are safe. A hiring decision does not add the expected capacity, so the people around the hire carry the gap through support, interruption, and repeated explanation. A founder or manager keeps the external story moving, so someone closer to the work absorbs the distance between that story and the reality of delivery.

None of this is unusual in small doses. Teams help each other. Senior people support juniors. Experienced people carry context. A startup cannot document, automate, or formalise everything from the beginning.

The pattern begins when the organisation starts depending on that extra human carry without naming it as cost.

On paper, the team may have enough people. In practice, the real capacity may be much smaller because senior attention is being consumed by support, repair, coordination, and interpretation. On paper, the system may have a process. In practice, the process may only work because a few people know how to make it work. On paper, a problem may be contained. In practice, containment may mean someone is spending part of themselves to stop it spreading.

Human debt is easy to miss because it enters the system through helpfulness.

Early Signals

The earliest sign is repeated quiet compensation.

The same person keeps filling the same gap. The same senior people keep being interrupted. The same context holders are pulled into decisions that should no longer require them. The same emotional stabilisers keep smoothing tension so the work can continue. No single moment looks unreasonable, but the repetition reveals where the system is outsourcing its weakness.

Invisible support work may also expand around visible work. The visible task may be a feature, a deployment, a report, or a new hire. The hidden work is the explanation, checking, reassurance, repair, translation, and follow-up needed to make that visible task function. If the hidden work keeps growing, the organisation may be accumulating human debt even while the visible plan appears staffed and sensible.

Knowledge becoming a living dependency is another sign. Certain people know the workaround, the customer history, the fragile service, the undocumented reason a decision was made, or the real meaning of a vague requirement. Their knowledge keeps the system moving, but it also reveals that the system has not made that knowledge shareable enough to survive without them.

The early signal is not exhaustion. That comes later. The early signal is reliance: the organisation keeps relying on human attention to compensate for structural gaps it has not yet resolved.

What Gets Misread

Human debt can look like loyalty.

The people carrying it may be capable, committed, and deeply invested. They may take pride in being useful. They may not complain early because the extra load feels like part of building something difficult. From the outside, their effort can look like culture working well.

But loyalty and load are not the same thing. A person can care about the company while carrying cost the company has failed to account for.

Headcount can create a second misread. Adding people can look like adding capacity, but people do not become capacity automatically. If the system is unclear, if the work is too dependent on local knowledge, or if the hires need more support than the organisation can afford, headcount can increase the visible team while reducing the practical capacity of the people under strain.

Resilience is another common misread. When the same people keep finding a way through, leaders may assume the system is more durable than it is. What they are really seeing may be personal elasticity. The product works because someone remembers. The release succeeds because someone watches it closely. The customer is reassured because someone absorbs the frustration. The system is not healthy. Human margin is being spent.

This is also where human debt differs from normal overwork. Overwork is visible in hours and load. Human debt can exist before the hours look extreme. It is the transfer of unresolved structural cost into people, even when the calendar looks manageable.

Hidden Cost

Human debt means the organisation has started financing its own continuity with people. Attention gets spent before it reaches the work that most needs it. Patience gets consumed by repeated explanation. Judgement gets narrowed by constant interruption. Trust wears down when people realise the same burden keeps returning to the same hands. The work may move, but the people carrying it have less room to think, recover, challenge, or care with the same openness they had before.

Unlike technical debt, human debt cannot be repaid by rewriting a module or cleaning up a backlog. Some of it can be reduced by better structure, clearer ownership, stronger process, and more honest staffing decisions. But once the cost has lived in people, it leaves marks that are not purely operational.

People change how available they are. They become more guarded with their effort. They ration care. They protect focus more aggressively. They may keep performing, but with less trust that the organisation understands what it has been taking from them.

That cost can also travel. A person who has been made into the buffer for one fragile system does not always leave that experience behind when the project ends. They may carry it into the next role as caution, lower trust, sharper boundaries, or a reduced willingness to give the next organisation the benefit of the doubt.

Human debt matters to the arc of the book because it is not a sentimental side effect. It is a structural consequence that changes the behaviour of the people the system still needs.

Recognition

Human debt becomes visible when you ask what the system is quietly asking people to absorb.

Look past the official workload. Look at who is interrupted, who explains, who remembers, who smooths tension, who catches the problem before it becomes visible, and who keeps carrying context that should have been made shareable by now.

Then look at where the plan assumes capacity that only exists because someone keeps compensating.

Human debt hides inside apparent sufficiency. The team looks staffed. The process looks defined. The project looks covered. But if the plan only works because a few people keep lending extra attention, patience, memory, or emotional steadiness, the organisation is not adequately resourced. It is borrowing from its people.

Teams need care. Startups need people who help each other across unclear edges. The danger begins when care becomes the substitute for structure.

Human debt accumulation is easy to miss because the work still gets done. The question is whether the system is genuinely holding, or whether people have become the place where its weakness is being stored.

Chapter 20
Normalised Overextension

When exceptional effort becomes ordinary capacity.

Normalised overextension begins when exceptional effort stops being treated as exceptional.

The stretch works, so it gets repeated. Someone stays late. Someone absorbs the extra coordination. Someone carries the awkward handover, the support issue, the customer escalation, or the work that should probably have been replanned. At first, everyone knows it is a stretch.

Then the stretch becomes familiar.

At that point, too much stops looking like too much. The company is no longer relying on an occasional hard push. It is treating personal stretch as ordinary capacity.

This is the movement from human debt into normalised load. Human debt accumulation describes where unresolved burden first gets stored. Normalised overextension describes what happens when carrying that burden stops feeling temporary and becomes part of how the organisation expects work to happen.

One project made that shift visible through a moment that should have changed the room.

By then the work had stopped feeling difficult in the usual way. It was no longer only a matter of solving hard problems. It had become a matter of enduring them. The team had spent months inside long hours, constant attention, slow feedback, and effort that did not seem to return enough progress for the amount of strain being spent.

No one needed to announce that morale was dropping. It was already in the atmosphere.

Then one day, one of the senior people exploded.

It was not staged or strategic. It was pressure escaping containment. Frustration that had clearly been building for a long time finally found a place to go.

The room went quiet. What stayed with me was not only the outburst, but my reaction to it: relief.

Thank God it was not me. Thank God my frustration had not surfaced first.

That reaction mattered because it told me the strain was not isolated inside the person who had broken visibly. It was shared. The outburst did not introduce a new fact into the room. It revealed something many people had been carrying privately.

The person took time away. When he came back, he apologised privately to people around the room. Everyone accepted it. Work continued. That was the part that mattered most.

The rupture did not become operating data. It did not cause the organisation to ask what level of strain had made that moment feel understandable. It was treated as something to move past.

After that, the atmosphere shifted in a subtle but lasting way. Tension became more ordinary. Irritation no longer stood out as sharply. People narrowed their focus just enough to get through the day. The project was alive. Commits were happening. From the outside, it probably looked like a team pushing through a hard phase.

Inside the room, something else had been absorbed.

The strain had stopped functioning as a warning.

How It Forms

Normalised overextension usually forms through repetition rather than command.

No one has to announce a new rule. The organisation learns it through what works. A late night gets the release out. A senior person absorbs another interruption and the project keeps moving. A team pushes through a rough week and the milestone is saved. The visible result arrives, so the system records the behaviour as available capacity.

The next time pressure rises, the same pattern is easier to repeat. The extra effort no longer feels like an exception. It feels like the known path through strain. People may notice the cost, but the noticing loses force because the company has proved it can keep going this way.

The baseline moves over time. A pace that would once have triggered concern becomes normal delivery. A person who once helped beyond their role becomes understood as the person who carries that category of problem. A team that once stretched for a deadline becomes expected to stretch for the next one as well.

Normalised overextension is difficult to interrupt because it does not always look like abuse or obvious mismanagement. It often looks like committed people doing what has already been shown to work.

Early Signals

Normalised overextension often begins with recovery disappearing from the operating model.

People may talk about taking time later, slowing down after the release, or cleaning things up once the pressure passes. But the plan does not actually contain recovery. The next push is forming. The next commitment is visible. Rest becomes something people are supposed to find privately, not something the system accounts for as part of delivery.

When that happens, recovery is often taken from whatever life still sits outside the work: sleep, family, relationships, health, and the quiet spaces where a person would normally become themselves again.

Visible strain may also stop creating alarm. Late nights no longer need explanation. Irritation blends into the background. A tense meeting is treated as normal for the phase. A visible outburst is managed as an interpersonal incident rather than read as evidence that the environment itself has become overloaded.

Sometimes the signal is quieter. In one long-running delivery effort, the clearest image was seeing the same developers still at their computers late in the evening, then arriving the next morning to find them in what looked like the same chairs, in the same positions, as if they had barely left. Whether they had stayed extremely late or returned very early almost did not matter. The effect was the same. The team had become part of the infrastructure keeping the system upright, and that was looking ordinary rather than alarming.

The pattern also shows up when new work is absorbed rather than exchanged. Meetings acknowledge that everyone is busy, but add more. Support demands increase, but the original delivery expectation stays intact. A person returns from leave and is immediately pulled back into the same pressure. The organisation may sympathise with the load while depending on it.

The strongest signal is baseline shift. Capacity is inferred from what people have already survived. Because the team got through the last push, the next plan quietly assumes it can get through another.

What Gets Misread

Normalised overextension can look like resilience.

The team delivers. Customers see progress. Leaders receive updates. Work continues to move. From a distance, the organisation can look strong, adaptable, and committed.

But resilience includes recovery. Overextension consumes it.

Culture is another misread. Some teams take pride in being intense, responsive, and willing to do what others will not. In a startup, that pride can be part of the early energy that makes difficult work possible. The risk begins when the company stops distinguishing between commitment and depletion. A team can care deeply while operating beyond what it can sustain.

Capacity is misread when endurance becomes planning evidence. If a person has repeatedly carried more than their role should require, the organisation may start treating that history as evidence of capacity rather than evidence of strain. What they endured becomes what is planned around. The fact that they kept going becomes a reason to ask again.

AI can sharpen this misread. If AI increases visible throughput, the new pace can quickly become the assumed pace, even when the surrounding human judgement, review, ownership, and recovery have not increased at all. The team may produce more artefacts, more code, more drafts, or more apparent progress, while the human system around that output operates with the same limited margin.

Hidden Cost

Normalised overextension makes margin disappear.

Margin is what allows people to think before reacting, challenge a weak assumption, help someone without losing their own work, recover from a hard week, or notice that a small issue is becoming a pattern. When margin disappears, ordinary problems start landing harder than they should.

A sick day becomes a delivery risk. A support spike derails planned work. A hard bug consumes people who are stretched. A staffing mismatch becomes personal strain instead of a solvable management issue. The organisation has less ability to absorb reality because it has spent the slack that would have made absorption possible.

The cost also changes behaviour. People become more reactive. Patience shortens. Challenge feels more expensive. Preventative work loses ground because everyone is focused on what is immediately in front of them. Small irritations become part of the atmosphere. The company may not yet be seeing emotional withdrawal, but it is creating the conditions for it.

Normalised overextension can last longer than it should because people can perform inside it for a long time. The danger sits there. The system keeps receiving output, so the cost remains easy to discount. But the output is being financed by depletion, narrowed attention, and recovery time that should never have been treated as ordinary capacity.

Recognition

Normalised overextension has a blunt test: does ordinary progress now depend on extraordinary stretch?

If the answer is yes, the organisation is not simply in a hard phase. It has changed the baseline. The question is no longer whether people are committed. The question is whether the company has started building its operating model around levels of effort that should have remained exceptional.

Recovery gives the second test. Is it treated as part of delivery, or as a luxury people can take later?

Recovery is not separate from capacity — it is one of the conditions that makes capacity real. A system that plans work without planning recovery is not being lean; it is borrowing against people.

Startups will always have periods of intensity. Urgency becomes dangerous when it stops being named, bounded, and paid back. Name the stretch. Remove or defer work when new work is added. Treat repeated late nights, recurring outbursts, support load, and disappearing recovery as operating signals, not personality quirks.

Normalised overextension is the point where exceptional effort starts being counted as ordinary capacity. The output may still arrive, but it is being financed by people with less margin than anyone has admitted.

Chapter 21
Emotional Withdrawal

When the truth remains in the room, but stops travelling.

Emotional withdrawal begins when people still see what is wrong, but stop bringing the full weight of that perception into the room.

They keep working. They keep noticing. They keep functioning. They may sound professional, committed, and calm. From above, the team can even appear more mature because the visible friction has reduced.

What changes is how much of themselves they allow the system to receive.

Signals are edited before they are voiced. Reactions are flattened into safer language or held back entirely. Questions are softened. Expertise is offered more selectively. Concerns move into side conversations, private messages, code comments, and one-to-ones instead of travelling cleanly into the forums where direction is set.

This is not the same as laziness or indifference. It is usually self-protection. After enough rounds of being ignored, overloaded, corrected too harshly, or asked to keep carrying the same unresolved strain, people learn to reduce the exposure that comes with full honesty.

Human debt describes the burden being stored in people. Normalised overextension describes that burden becoming baseline. Emotional withdrawal describes what happens when people remain present, but begin bringing less signal into the formal life of the system.

I saw this most clearly in one project, through the difference between two rooms.

People close to the work could see the widening gap between the formal story and the lived condition of the system. The product was unstable. Workarounds were multiplying. The effort required to keep things moving was becoming harder to explain from the outside because so much of it lived in local knowledge, patches, and people quietly holding pieces together.

Then, in one internal room, the truth came out more plainly. Engineers, testers, and the people closest to the day-to-day delivery spoke about the instability, the shortcuts, the fragile paths through the system, and the sheer amount of effort required to keep the project upright. The room was not easy, but it was honest. People were not performing confidence. They were naming what the work had become.

For a moment, that honesty felt useful.

There was relief in it. Not because the problems had been solved, but because they had finally been spoken in a room that seemed capable of receiving them. The strain had moved from private knowing into shared language. That alone can feel like movement when people have been carrying something for too long.

I remember the feeling that the truth might now travel. Not perfectly. Not magically. But enough that the organisation might pause, regroup, and make room for a more honest conversation about recovery.

Maybe the organisation would pause. Maybe recovery would be discussed seriously. Maybe the gap between the formal story and the lived system had become visible enough that it could no longer be softened away.

But in the later formal settings, the truth did not arrive with the same weight.

The tone returned to something cleaner. Progress was acknowledged. People were recognised for pushing through. The project was spoken about in language that made the situation sound difficult but manageable. What had been heavy in the internal room came back as a lighter story.

That was the lesson: the truth could be spoken in one room and still return changed in another.

After that, meetings became something to survive. People saw the problems. They talked about them in smaller groups, side conversations, and technical spaces. But less of the full signal travelled into the rooms where direction was set. The organisation had not silenced anyone in an obvious way. It had simply taught people that formal escalation might not carry the gravity of what they knew.

The quiet turn in emotional withdrawal is not that people stop caring. They stop believing the system will do anything useful with the full truth.

How It Forms

Emotional withdrawal usually forms when truth stops producing useful movement.

A concern is raised and acknowledged, but nothing changes. A difficult issue is softened as it travels upward. A person names a risk and inherits more work without more authority. A half-formed question is exposed too harshly in a meeting. A team speaks honestly in one room, then watches the next formal setting return to the safer story.

Each moment teaches a small lesson. Speaking plainly may cost energy. Surfacing risk may create friction without creating action. Asking the harder question may make the person asking it look negative, difficult, or slow. Over time, people adjust before they speak.

The adjustment is rarely dramatic. People do not announce that they are withdrawing. They simply become more careful. They wait longer before raising a concern. They translate sharper reactions into softer language. They stop volunteering unfinished thoughts. They bring the polished version of their expertise, not the full version.

The room can function inside this condition. Meetings happen. Work progresses. People may be polite, constructive, and technically capable. What has changed is the quality of signal the organisation is receiving.

Early Signals

Emotional withdrawal can start as calmer rooms with poorer truth.

The discussion feels smoother because less is being contested. Fewer difficult questions interrupt the plan. Fewer people visibly react to weak assumptions. The meeting may feel aligned, but the harder conversation has moved elsewhere.

Self-editing is another clue. People know what they think, but they begin trimming it before the room receives it. A serious concern becomes a careful caveat. A strong technical objection becomes a mild risk note. A product contradiction becomes something to discuss offline. A person who would once have pushed for clarity now asks a narrower, safer question.

The after-meeting tells the truth. The real conversation happens once the formal room has ended. People gather in smaller groups and say what they did not say earlier. They compare doubts, frustration, and recognition in places where the organisation cannot fully use them. The truth is still alive, but it is no longer travelling through the channels that shape decisions.

Selective documentation can carry the same signal. Issues stay in code comments, local notes, private memory, or informal handovers rather than entering the systems that would force visibility. The people closest to the work see the risk. They have simply stopped believing that formal escalation will help.

What Gets Misread

Emotional withdrawal can look like alignment.

When challenge reduces, leaders may assume the team has accepted the plan. When meetings become calmer, they may assume the conflict has been resolved. When fewer concerns are raised, they may assume the concerns have gone away.

Sometimes that is true. Often, it is not. The concern has not disappeared. It has become more private.

Professionalism can also hide the pattern. People under strain learn to manage their tone carefully. They become measured, neutral, and polite because that is the safest way to keep functioning. From a distance, that can look like maturity. Inside the team, it may feel more like survival.

Busyness is another way the withdrawal stays hidden. People can remain busy, engaged, and technically stimulated while withdrawing from product truth. A team may keep building, learning, solving, and shipping while avoiding the harder question the organisation no longer seems willing to answer. Motion can mask withdrawal because the work continues even as the most important signal weakens.

Emotional withdrawal is dangerous in technology startups because it does not always make the room obviously silent. It makes the room less complete.

Hidden Cost

Emotional withdrawal reduces the organisation's ability to sense itself. The system loses access to part of what its own people know. Urgency becomes caution. Concern becomes a muted note. Contradiction becomes private conversation. Expertise becomes conditional. The organisation may receive updates, but it is no longer receiving the full condition of the work.

Trust also thins in a quieter way. When one room can hold the heavier truth and the formal room returns a lighter version of it, people stop assuming the organisation is carrying the same reality they are. They begin guessing what others really believe, what will survive translation, and whether full honesty will simply come back as a cleaner story. That uncertainty does not only weaken candour. It weakens shared confidence that the system can still tell the truth to itself.

One consequence is the shift from prevention to reaction. The people closest to the build usually know where the fragility is, where the product is likely to fail, and which issues are waiting in the grass. But after enough rounds of not being heard, being softened, being overloaded, or feeling unsafe, they stop believing it is worth raising those issues early and formally.

Less gets documented. Less gets challenged. Less preventative work is protected. The same team that might once have surfaced risk early begins spending more effort fixing what has already escaped.

That cost can feed earlier patterns back into the system. Process erosion deepens because less gets raised or recorded in formal systems. Borrowed confidence strengthens because fewer people push back on the reassuring story. Incentive drift becomes easier because the work that looks good faces less challenge from the people who can see what it is displacing.

From altitude, later bugs, unstable releases, and strange customer behaviour can look sudden. Often they are not sudden at all. The people closest to the work knew first. What changed is that less of that knowledge travelled upward clearly enough, early enough, or formally enough to alter direction before the cost reached customers and the wider business.

Recognition

Emotional withdrawal is not measured only by whether people are still speaking.

The harder question is whether they still feel safe bringing the full weight of their expertise.

The distinction matters. A team can keep talking while withholding the sharper truth. It can keep attending meetings while shifting real concern into private channels. It can keep delivering work while no longer believing the formal system will use what the team knows.

Then ask where the truth goes after the meeting.

If the strongest version of reality appears only in side conversations, private messages, code comments, or one-to-ones, the organisation has a signal problem. Demanding more candour as a performance expectation will not solve it. The better question is what the system has taught people about the cost and usefulness of speaking plainly.

Endless emotional processing is not the answer either. The work is to make truth useful again. When people surface risk, something must happen with it. When a concern is raised, it must not return as a lighter story. When someone brings uncertainty into the room, the room must treat that uncertainty as material for better judgement rather than as a personal flaw.

Quiet is the misleading signal. Sometimes the room is calmer because people trust the system. Sometimes it is calmer because they have learned to protect themselves from it.

Chapter 22
From Human Cost to Reckoning

Late in the evening a meeting finally ends.

Most of the discussion sounded reasonable. The roadmap was reviewed. Deadlines were adjusted slightly. A few risks were acknowledged but set aside for later investigation. From the outside, the organisation still appears to be operating normally.

When the call ends, several people remain quietly at their desks.

One person closes their laptop and exhales, aware that tomorrow's work will begin with problems no one quite named in the meeting. Another begins writing a private note about an issue that feels important but difficult to raise again. A third simply leaves the office without saying anything at all.

Nothing dramatic has happened.

But something has changed.

The system is still moving forward, yet fewer people believe it can correct itself in time.

Part IV described what happens when the strain created during accumulation moves into people.

Human debt accumulation transfers unresolved load into attention, patience, memory, judgement, care, and endurance. Normalised overextension turns that stretch into the expected baseline of participation. Emotional withdrawal filters the signals that once helped organisations sense when something was going wrong.

Taken together, these are not three unrelated harms. They describe a sequence inside the human layer of the system: burden is first absorbed, then planned around, then less fully spoken.

Each adaptation preserves stability in the short term. The work continues. Delivery remains possible. Meetings still end with decisions.

But these adaptations also remove the very mechanisms that once allowed the system to recognise its own limits.

They do not merely accompany the arc.

They help carry it forward.

Trust becomes thinner. Recovery disappears. Signals travel more slowly. Concerns surface later, and often more softly than they once would have.

From the outside the organisation still appears functional.

Internally the system has become fragile.

What has accumulated across these chapters is not simply pressure, but distance: distance between what people are experiencing and what the organisation is able to acknowledge.

For a time that distance can be maintained. People compensate. Stretch becomes normal. Truth travels through softer channels. Momentum continues.

Eventually, however, the system encounters something it cannot quietly absorb.

A decision exposes the gap between plans and reality. An incident reveals how much strain the organisation was already carrying. A moment arrives when the distance between signal and response collapses.

That moment is the reckoning.

Part V begins where the system can no longer maintain the distance that allowed it to keep moving.

If you want to understand why that confrontation so often feels sudden at leadership altitude, look at what people have been quietly carrying in order to keep the system feeling stable.

Part V

The Reckoning

"The bill does not suddenly appear.
It reaches the point where it must finally be paid."

Chapter 23
When Deferral Turns Visible

The reckoning is not the moment everything fails.

It is the point at which deferral becomes visible.

For a long time, a company can move with consequences held at a distance. Trade-offs are made under pressure. Questions are postponed because the answer would slow the work. Human strain is absorbed quietly. Technical debt is worked around. Process gaps are covered by memory, goodwill, and effort.

At the time, much of this feels practical. A startup is always deciding what it can afford to face now and what it has to carry forward.

The reckoning begins when the organisation can no longer keep that separation intact.

Costs that were once deferred become payable. Consequences that used to arrive slowly begin arriving closer together. Choices that still sounded open become constrained by what has happened. Ambiguity that once helped the system keep moving starts creating friction, exposure, and responsibility.

Reckoning often feels sudden even when it has been forming for a long time.

The visible moment is rarely the whole event. A failed launch, an uncomfortable board conversation, a data reconciliation, a customer escalation, a painful audit, a funding constraint, or a team exhaustion point may appear to be the beginning. More often it is the moment the system stops being able to conceal what earlier decisions created.

Part V enters that terrain.

The Bill Comes Due looks at the point where accumulated cost becomes concrete enough that someone has to pay attention to it. Compression of Consequence looks at what happens when the delay between reality and consequence collapses. The Narrowing of Options looks at the loss of meaningful flexibility before the organisation has fully admitted that the field has shrunk. Forced Clarity looks at the moment ambiguity stops carrying the system and reality becomes legible enough that ownership can no longer remain vague.

These are not four different names for failure.

They are four ways deferral runs out.

The distinction matters because reckoning can still become more than collapse. A system that sees the bill early may still change what it funds. A team that notices compression may still slow down enough to think. Leaders who recognise narrowing may stop pretending every option is still real. An organisation that can tolerate forced clarity may use it as the beginning of repair rather than treating visibility as attack.

The patterns in this part are therefore not written as predictions of doom. They are written as recognition points.

When a reckoning becomes visible, optimism alone is rarely enough. The margins are thinner. The buffer has been spent. The human system may be tired. The technical system may be brittle. The strategic field may be smaller than the language of the room suggests.

But visible does not mean finished.

It means the organisation is being asked to relate to reality more honestly than it did before.

What follows is not determined by intention, effort, or belief alone. It depends on whether the system can now confront what it has been avoiding, and whether the people inside it can tell the difference between a difficult truth and an unacceptable one.

Chapter 24
The Bill Comes Due

When scattered costs become unavoidable.

The bill comes due when costs that have been distributed across teams, workarounds, delay, and optimism finally condense into something concrete enough that no one can honestly call it temporary anymore.

A rewrite.

Another year.

Another million dollars.

No guarantee.

The bill is not always financial, but it usually becomes material. It becomes a decision the organisation must now face, a cost it can no longer spread thinly across people and time, or a repair too large to describe as cleanup.

Nothing about that moment is random. It is built out of earlier decisions that remained tolerable while their cost stayed scattered. The bill does not suddenly appear. It reaches the point where it must finally be paid.

This is the first hard movement inside Reckoning. Earlier chapters showed strain accumulating, landing in people, and weakening the system's ability to sense itself. Here, that strain becomes visible as consequence. Not collapse necessarily. Not punishment. A bill.

One indoor wayfinding project reached that point after a long period in which the cost had stayed distributed.

The product had a compelling premise and serious backing. It was meant to help people move through large, complex indoor spaces with more confidence. Over time, though, decisions accumulated around the product in ways that made it harder to keep coherent. A messaging layer was imposed where lighter options had once been available.

Reporting infrastructure was added through extra transformation paths. Build and release machinery became slower. The positioning system itself remained tied to real-world infrastructure that was never quite reliable enough for the experience the product implied.

For a while, none of that looked like one bill.

It looked like latency here, integration weight there, another awkward build, another workaround, another support conversation, another meeting where the team had to explain why a simple-sounding change was not simple. The cost was spread across time, attention, confidence, and technical friction. The project looked alive. The platform looked close enough to keep going.

Then the cost condensed.

To make the system reliable enough in the real world, major parts of the underlying hardware and network infrastructure would need to be redone. This was not a local fix or a little cleanup. It was another substantial investment, with no guarantee that paying it would fully solve the problem. That was the bill.

The number mattered, but the number was not the whole point. What mattered was that the organisation could not spread the cost thinly enough to keep calling it progress. The earlier strain had become a funding decision. The accumulated compromises, integrations, assumptions, and reliability gaps had gathered into something concrete enough that optimism could not keep it abstract.

The problem had not suddenly appeared.

It had simply reached the point where it had to be paid.

How It Forms

The bill forms while the cost remains distributed.

At first, a deferred fix remains small enough to carry. A workaround holds. A missing capability can be handled manually. An architectural constraint can be lived with. A fragile process can be managed by people who know the system well enough. None of these feels like a bill yet because none of them requires a single decisive payment.

The cost is spread across slower delivery, support effort, rework, confidence, recovery time, and the attention of people close to the work. It shows up as friction rather than invoice. That makes it easier to keep moving.

The system asks for more over time. A fix that once looked local now touches multiple layers. A product limitation that once looked acceptable begins constraining customer use. A technical choice that once seemed pragmatic now prevents the next version from being built cleanly. A staffing or vendor decision that once looked economical now requires rescue, translation, replacement, or rebuilding.

Eventually, the cost gathers into a form the organisation has to recognise. It needs a migration. A rebuild. A stabilisation phase. A support model. A new platform. A year of rework. A funding decision. A pause. The language changes because the condition has changed. The problem has stopped being carried invisibly. It is now asking to be paid.

Early Signals

The bill starts becoming visible when repair begins needing a number.

The organisation stops saying, "We should clean that up later." It starts asking how many months, how much money, how many people, how much downtime, or how much disruption it would take to make the system reliable enough. Once the question needs a number, the bill is forming.

Progress may also start requiring more explanation than movement. Tasks that once moved through the system now need coordination across teams, vendors, integrations, customers, or legacy decisions. People spend more time explaining why work is difficult than doing the work itself. The resistance has become part of the structure.

The bill is also forming when reversibility disappears. Earlier choices that once felt adjustable carry visible risk. People hesitate before changing them because the dependencies around them have grown. The organisation is making decisions, but the old sense of flexibility has thinned.

No-guarantee recovery is another sign. The company can identify a possible repair path, but cannot promise that paying the cost will restore the opportunity fully. The bill may be large, but the receipt is uncertain.

What Gets Misread

The bill can look like a sudden problem.

From leadership altitude, it can feel as if a new issue has appeared. A technical review returns an uncomfortable estimate. A customer condition exposes an old assumption. A rebuild becomes unavoidable. A system that looked close now requires another major investment.

But usually the bill is not new. What is new is its visibility.

A later bill can make earlier people look careless. Sometimes poor judgement is part of the story. More commonly, the bill has been assembled from reasonable decisions made under pressure: move faster here, defer that edge case, accept this constraint, preserve the demo, keep the customer warm, avoid slowing the team. Each decision may have been defensible. The bill arrives when the sequence cannot be defended as temporary.

Paying the bill can create its own misread. Sometimes it does put the company back in control. But paying for a rebuild, a migration, or a rescue effort does not automatically restore lost time, trust, customer confidence, or organisational energy. Some bills can be paid directly. Others are paid while the company also absorbs the fact that the earlier opportunity has changed.

This is what separates The Bill Comes Due from the chapters that follow. This chapter is about consequence becoming concrete. Compression of Consequence is about the delay between reality and consequence shortening. The Narrowing of Options is about what happens after earlier decisions have reduced the meaningful field of choice.

Hidden Cost

When the bill comes due, the company shifts from shaping outcomes to negotiating with reality.

Before the bill arrives, the organisation can talk in the language of possibility. It can frame work as sequencing, cleanup, iteration, productisation, or the next push. Once the bill is visible, that language becomes harder to sustain. The company is not only deciding what it wants to build. It is negotiating with what earlier decisions have made true.

Momentum slows because attention moves from creation to recovery. Confidence becomes more cautious because the cost is not theoretical. Teams spend increasing amounts of time stabilising what exists rather than building what comes next. Leaders experience decisions as constrained by the past.

The bill also changes the emotional climate. Optimism that once felt generative can feel expensive. People who warned earlier may feel vindicated but tired. People who believed the trajectory may feel exposed. The organisation has to decide not only whether it can pay the cost, but whether it can tolerate what the cost reveals.

That revelation matters. A bill is not just a price — it is evidence. It shows what was actually being carried, what was deferred, and what the system now requires before the story can continue.

Recognition

You can see the bill in costs the company treats as scattered after they have started becoming payable.

Look for the work that keeps returning with a larger estimate. Look for the area everyone knows is fragile but no one can fix locally anymore. Look for the deferred decision that now needs funding, staffing, time, or executive attention. Look for the problem whose repair path includes the phrase "no guarantee."

A harder test follows: what would have to be true for this to become ordinary cleanup again?

If the answer is unclear, the bill may have crossed into reckoning. That does not mean the company is doomed. It means the organisation has to stop treating the cost as future housekeeping and start treating it as present consequence.

Panic does not help here. Accounting does. Name the bill cleanly. Name what created it. Name what paying it can and cannot restore. Name what becomes impossible if it is not paid.

The bill comes due when the system cannot spread cost thinly enough to keep calling it progress. What had been deferred during accumulation is now visible as consequence. The organisation is not negotiating with possibility. It is negotiating with reality.

Chapter 25 Compression of Consequence

When delay stops protecting the organisation from itself.

Compression of consequence is what happens when delay stops protecting the organisation from itself.

For a long time, the costs of a struggling system can remain distributed across code comments, workarounds, longer days, private concern, and status reports that never carry the full weight of what people close to the work know. The same reality exists, but it does not reach everyone at the same speed.

Then the insulation thins.

What had been spread across weeks or months begins landing in days, meetings, customer interactions, and decisions. The underlying system has not suddenly changed. What has changed is the delay between that reality and its consequences.

This separates compression of consequence from The Bill Comes Due. The bill is about cost becoming concrete. Compression of consequence is about tempo. It is the moment the organisation loses the time lag that used to soften, distribute, reinterpret, or delay what the system was trying to say.

The opening chapter of this book began inside that kind of compression.

A senior business leader stood at the front of a room congratulating the team on progress, momentum, and the way people had pushed through difficulty. From his altitude, the project still looked like something that was moving forward. From the seats occupied by the people closest to the work, the reality felt very different.

The moment was not dramatic because new information had suddenly appeared. It was dramatic because old information had finally reached the decision room all at once.

People close to the work had been living with the strain for months. They had seen the performance problems, the workarounds, the growing shortcuts, the technical risks, and the quiet effort required to keep the product upright.

In one smaller internal room, some of that truth had already been spoken more plainly. The signs were not abstract. They were sitting in the codebase as comments, fixes, workarounds, and little markers of discomfort: notes to come back later, temporary patches, places where the system had been held together just long enough to keep moving.

The truth had not been absent.

It had been local.

But as it travelled upward, it kept returning lighter. What had been spoken plainly in smaller rooms was softened, filtered, or left behind in the formal ones.

That was emotional withdrawal.

By the time the senior leader stood in front of the wider room, there was no explanation. He read the reality in people's faces.

That was compression of consequence.

By then, the delay had collapsed.

The project did not become fragile in that meeting. It had been fragile before he arrived. What changed was the speed at which that reality reached leadership altitude. For the team, consequence had been approaching gradually. For him, it arrived almost at once.

Within days, the project was over.

Compression can therefore feel violent even when the underlying reality has been forming slowly. The people closest to the work experience a long descent. The people insulated from the work experience a cliff.

A similar pattern can appear outside leadership rooms. In another company, a booking platform that looked smooth on a carefully prepared sales laptop behaved very differently when it reached remote golf courses with slow, unreliable connections. In the demo environment, the gap could be softened by explanation. In the field, staff and customers felt it immediately while they waited for the system to catch up. The consequence had not been invented by deployment. Deployment removed the delay.

How It Forms

Compression forms when the buffers between reality and consequence are used up.

Earlier in the arc, weak signals can remain local. A fragile workaround stays inside the team. A customer problem is managed by a support person who knows the history. A concern is softened as it travels upward. A product gap is hidden by a demo, a manual process, or a patient insider who knows how to keep things moving.

Those buffers create delay. They are not always dishonest. Often they are simply the ordinary friction of organisation: reporting layers, interpretation, optimism, distance from the work, and the fact that not every problem needs to be escalated immediately.

But when strain accumulates, those buffers fail. The local workaround cannot stay local. The customer environment answers the assumption directly. The leadership room sees in one meeting what the team has lived with for months. A defect no longer waits for the next review cycle. A decision that once had room around it now produces consequence almost immediately.

The pattern is not that reality changes faster. It is that the buffers that once delayed its consequences have thinned, leaving the organisation less time to respond.

Early Signals

Compression often begins as the loss of breathing room.

Questions that once could remain open now demand quick answers. Risks that once sat in the background start shaping the next meeting. A weakness that might once have stayed inside one team begins surfacing across customers, leaders, or adjacent functions before anyone has time to package it neatly.

The gap between informal recognition and formal consequence also starts shrinking. People close to the work may have known the truth for a long time, but now the distance between what they know and what leadership must act on starts narrowing. What was privately understood begins arriving in places where it can no longer be treated as background noise.

Compression is also visible when the environment starts answering assumptions faster than the organisation can explain them. A demo that worked in controlled conditions meets a real customer environment. A platform that looked stable under internal use meets operational load. A process that relied on local knowledge meets a handover, absence, or support spike. The consequence does not wait for a retrospective. It appears in use.

The strongest signal is speed. The same issue that once would have taken weeks to surface now lands in days. The same concern that once could be softened now arrives with less room for reinterpretation.

What Gets Misread

Compression of consequence can look like sudden failure.

From one altitude, it can feel sudden. A senior leader walks into the room and reads the strain all at once. A customer encounters the product in real conditions and the weakness becomes visible immediately. A project that appeared to be struggling but manageable turns into a decision problem within days.

From another altitude, it was not sudden at all. The team had been living inside the reality for a long time. The consequence arrived quickly only for the people who had been insulated from it.

The triggering event often gets too much credit. A meeting, customer deployment, audit, or executive review may appear to create the crisis. Usually it exposes the crisis. The event matters because it removes the delay, not because it invented the condition.

Urgency can also be mistaken for clarity. When consequence compresses, decisions feel more urgent and the organisation may mistake that urgency for sharper understanding. But speed can narrow interpretation. People respond to the pressure of consequence before they have fully understood the sequence that produced it.

Compression is dangerous because it does not only reveal consequence. It changes the conditions under which the organisation must interpret consequence.

Hidden Cost

Compressed consequence narrows mental and organisational space.

Reflection shortens. Recovery windows shrink. Leaders become more cautious. Teams begin testing ideas socially before risking them operationally. Proposals are softened, delayed, or abandoned earlier because everyone can feel how little time remains between action and fallout.

This can look like prudence, and sometimes it is. A system that has lost margin should behave with more care. But compression also means the organisation is now making decisions under reduced interpretive space. It has less time to separate symptom from cause, less room to test alternatives, and less capacity to absorb the emotional force of what has become visible.

The cost also lands unevenly. People close to the work may feel exhausted because the visible consequence confirms what they have carried. People further away may feel shocked because the recognition has arrived all at once. Those two experiences can collide. One group feels, "We have been saying this." Another feels, "Why did no one tell us clearly enough?"

That collision can damage trust unless the organisation understands the pattern. Compression reveals not only the technical or operational issue, but the different speeds at which reality has been moving through the company.

Recognition

Compression often starts where the organisation has been relying on delay to make reality feel manageable.

Look for problems that stay tolerable only because they surface slowly. Look for weak signals that remain local until the right customer, leader, deployment, or operational condition removes the buffer. Look for issues that people close to the work understand but that have not yet reached the places where formal decisions are made.

It also helps to ask who has been living with this reality longer than leadership has been recognising it.

Compression often creates false surprise. The organisation experiences the consequence as new because the recognition is new, not because the reality is new.

Some delay is necessary. Not every signal should become an emergency. What matters is whether the organisation has visibility loops strong enough for important reality to travel before crisis conditions force it into the room.

Compression of consequence turns distance into immediacy. What had felt survivable while distributed becomes harder to interpret because it is no longer arriving one piece at a time.

Chapter 26
The Narrowing of Options

When choices remain visible after real options have already shrunk.

The narrowing of options begins long before anyone says the company is running out of choices.

Architecture does it. Budget does it. Procurement does it. Repeated compromise does it. Time does it. So do confidence, reputation, staffing, customer promises, and the need to keep a story alive after the system has started pushing back.

Meetings happen. Alternatives are named. People can speak in the language of strategy, trade-offs, and next steps.

But the meaningful field has started to shrink.

The pattern is difficult to recognise because the organisation can feel as if it is choosing freely because it is discussing options. In reality, many of those options have lost force. They remain available in theory, but not at a cost, speed, or level of disruption the company can carry.

The Bill Comes Due shows cost becoming concrete. Compression of Consequence shows the time lag collapsing. The Narrowing of Options shows what happens when earlier decisions have removed many of the futures the organisation speaks about as if they are open.

One indoor wayfinding project showed this pattern clearly.

The product was ambitious in a way that made sense. Large campuses are difficult to navigate. Buildings have levels, corridors, lecture theatres, meeting rooms, cafes, libraries, and hidden routes that people learn slowly if they learn them at all. A good indoor navigation product could help people find their way and, over time, help the organisation understand how space was actually being used.

The more impressive version of that future was highly automated. The app would know where a person was. It would infer movement through the building. It would guide them from place to place with the confidence people had come to expect from outdoor maps. That version was attractive because it sounded like the real future of the product. It felt more complete, more polished, and more platform-like.

But there was a simpler path available.

One person close to the work had built and argued for a lower-complexity wayfinding approach. It accepted more manual input from the user. Instead of requiring the system to solve precise indoor positioning from the beginning, it could let the user confirm where they were and then guide them from there. It traded sophistication for reliability. It was less magical, but more achievable.

That kind of option can be hard for an ambitious project to choose.

It asks the organisation to accept a smaller version of the promise. It makes the product less automatic. It requires people to admit that the impressive path may not be the most useful first path. In this case, the more automated direction kept more force. The simpler route remained technically imaginable, but it did not become the path the project organised itself around.

From there, other decisions began to narrow the field further.

Enterprise tools were brought into the architecture. A proprietary messaging layer replaced lighter options that had been working well enough. Reporting and analytics were retrofitted through products that made organisational sense but added integration weight. Centralised build pipelines slowed development. The positioning approach itself depended on Wi-Fi signal accuracy that was never reliable enough to support the experience the product implied.

Each decision could be explained. The tools were serious. The reporting mattered. The platform needed to look credible. The organisation had existing systems, existing procurement logic, and existing expectations about what an enterprise product should use.

But together, those choices changed the decision field.

Once the product's limitations became difficult to ignore, the company was not choosing among the same set of futures it had once had. A simpler, more practical path had existed earlier, but the project had moved away from it. The more automated path had consumed time, money, attention, and credibility. The remaining alternatives now required infrastructure change, renewed funding, and a willingness to unwind decisions that had become part of the project's identity.

The option had not disappeared in a literal sense. It had become organisationally unavailable.

This is one way options narrow: the company still wants progress, but no longer wants to pay the structural cost required to make that progress possible.

How It Forms

Options narrow when decisions create dependencies that later decisions must respect.

At first, those dependencies can look harmless. A cheaper architecture is chosen to preserve momentum. A platform is adopted because it seems serious, available, or paid for. A more ambitious path wins because it looks more impressive than the simpler one. A rebuild starts before the organisation has absorbed why the last version failed. A temporary workaround becomes the thing other work is built around.

These choices usually form under real constraints: money, time, credibility, customer expectation, investor pressure, and the need to keep visible progress alive. The narrowing happens because every decision changes the cost of later movement.

The longer the company waits, the less neutral waiting becomes. A choice deferred today may exist tomorrow, but tomorrow it may require more money, more rework, more authority, more trust, or more appetite for disruption. Sometimes the option has not disappeared. It has simply become too expensive for the organisation to treat as real.

The company begins navigating inside a box built from its own prior decisions. It may ask, "What should we do?" but the more honest question has become, "What can we still do from here?"

Early Signals

Options start narrowing when proposals arrive with disclaimers attached from the beginning.

People stop describing the best path. They explain why the best path is unlikely to be acceptable. They pre-empt the budget objection, the schedule objection, the procurement objection, the customer objection, or the political objection before the room has even considered the idea. The option is named, but it arrives weakened.

Conversations also shift from exploration to mitigation. The question moves from "What would solve this properly?" to "What can we carry now?" Teams stop discussing the broader set of possible futures and focus instead on the least damaging path through the current constraint.

Narrowing is also visible when simpler paths lose internal force. A practical, lower-complexity option may be technically possible, but the organisation has organised itself around the more ambitious, more automated, more enterprise-looking, or more impressive direction. Going back to the simpler path would now feel like retreat, even if it might have created more usable value sooner.

People close to the work may start dismissing options quickly because they know the hidden cost. From the outside, this can look like negativity. From the inside, it may be accumulated knowledge: everyone who understands the system knows what that option would require.

What Gets Misread

The narrowing of options can look like planning.

The organisation is holding meetings. It is comparing paths. It is making decisions. That activity can create the impression that the strategic field remains open.

But planning and optionality are not the same thing. A company can plan carefully inside a field that has shrunk.

Realism can carry the second misread. When people say, "We have to be realistic," they may be doing important discipline. Startups need realism. But realism can also become the language used after earlier decisions have quietly removed better futures. The company may present the remaining path as prudent when it is really constrained.

Narrowing can also be mistaken for failure having happened. It has not. A narrower field can contain viable paths. The danger is pretending the field is wider than it is. When leaders believe they are choosing among broad strategic alternatives, but the system is actually offering only constrained responses, decisions get made against a false sense of freedom.

This pattern is especially visible in technical products because architecture preserves history. Code, infrastructure, data shape, deployment model, vendor commitments, integration choices, and product positioning all carry earlier assumptions forward. When a new decision is required, the system may know which futures it can and cannot support.

Hidden Cost

The narrowing of options costs the organisation meaningful flexibility before it has fully acknowledged the loss.

This cost is not only strategic. It changes the emotional and political shape of decision-making. People become more cautious about naming the clean answer because the clean answer is not practical. Teams propose smaller moves because larger ones would expose how much has been consumed. Leaders keep asking for alternatives, but the people closest to the work know that many alternatives are now mostly ceremonial.

The company can move, and that matters. The narrowing of options is not immobility — it is constrained movement. The organisation can choose, but its choices increasingly carry the shape of what has already happened.

The pattern can feel frustrating rather than dramatic because no one can point to the single moment the options disappeared. They disappeared gradually, through decisions that each made sense at the time. Once the narrowing is visible, the organisation may be left deciding among partial recoveries, compromised futures, and expensive corrections.

The deeper cost is that possibility starts being confused with availability. An option can be technically possible without being organisationally available. It may require funding the company lacks, trust it has spent, time the market will not give, or a level of disruption the team cannot absorb.

Recognition

The useful test is to separate the options that are still real from the ones that are only being named.

That distinction cuts through a lot of meeting theatre. A real option has a plausible path to action. It has someone willing to own it, enough resources to pursue it, and a cost the organisation is prepared to face. A named option without those things may be useful for thinking, but it is not genuine strategic freedom.

In technical systems, this distinction is especially visible. In a farm-control product built around a latency-tolerant model, a more responsive architecture was still possible after the mismatch became visible. But it did not resemble an ordinary design choice. It meant revisiting control paths, timing assumptions, and months of code history. The option was technically possible. It was not organisationally available in the way it would have been earlier.

Then ask what waiting has changed.

If waiting has made the clean path more expensive, the simple path less credible, the repair path more disruptive, or the customer promise harder to unwind, then waiting has narrowed the field. The organisation may not have made a new decision, but the decision environment has changed around it.

Startups have to commit. Keeping every option alive is impossible. What matters is knowing which commitments are reducing reversibility and refusing to pretend that all later choices remain equally available.

By the time the narrowing is obvious, strategy may still sound wide open while the organisation is already acting inside a box built by earlier decisions. What used to feel like navigation becomes selection among constraints. The meaningful field has shrunk.

Chapter 27
Forced Clarity

When ambiguity can no longer carry the system.

Forced clarity is the moment when ambiguity stops carrying the system.

For a while, a company can keep moving on partial understanding. It can rely on local knowledge, manual handoffs, incomplete reports, founder memory, spreadsheet logic, and people who each understand one part of the flow well enough to keep their own area moving.

That arrangement can be inefficient, but it can still be survivable. The business does not need to see itself clearly as long as the gaps remain local, the costs remain absorbed, and no one has to reconcile the whole picture at once.

Then something changes. A report works. An audit lands. A rebuild exposes the logic. A simple question travels further than expected. A customer issue refuses to stay isolated. A meeting that was supposed to close a point opens a wider contradiction.

Clarity arrives not because the organisation suddenly becomes wiser, but because it runs out of ways to remain unclear.

The Bill Comes Due shows cost becoming payable. Compression of Consequence shows the time lag collapsing. The Narrowing of Options shows the field of viable choices shrinking. Forced Clarity shows the moment the system becomes legible enough that ambiguity cannot protect it.

One fast-growing service business made this pattern visible through a reporting project that started as a technical problem and became something larger.

The company was acquiring customers quickly through introductory offers and discounted pricing. The business had serious financial backing, but locally it carried the energy and disorder of a company building itself while in motion.

The systems underneath had not kept pace.

Customer information, billing inputs, pricing data, cost calculations, and operational assumptions lived across a purchased CRM, spreadsheets, manual handoffs, and the heads of subject-matter experts. Many people understood their own part of the process well. One person knew a smaller group of closely managed accounts. Another knew a particular spreadsheet. Someone else knew how a set of cost inputs was being handled.

Almost no one understood the whole thing end to end.

That was survivable while the business could keep treating the system as a set of local responsibilities. The CRM carried part of the truth. Spreadsheets carried other parts. People filled gaps with memory, experience, and workarounds. It was messy, but the mess had not yet forced itself into one shared picture.

The work was to build a reporting bridge around that fragmented environment. The team pulled together data from the CRM, spreadsheets, databases, and other operational sources so the business could produce something like a meaningful cost-of-goods-sold report. It took months of tracing fields, asking what each spreadsheet meant, and working out how one person's output became another person's input.

When the reporting path finally worked, the first reaction was promising. Some lines made sense. A senior finance person could look at particular figures and say they felt right. The system was becoming legible enough to test.

Then the picture worsened.

The lines that made sense were exceptions. The smaller, more closely managed part of the customer base was easier to validate. In the high-volume segment, where speed and scale had mattered most, the results deteriorated quickly.

That was the forced-clarity moment.

The reporting bridge was not failing because the logic was unclear. It was working well enough to expose that the underlying records were wrong. Customer pricing data had been entered incorrectly. Cost inputs were unreliable. The business had not merely lacked a report. It had been feeding inaccurate information into the system it needed to trust.

Once that became visible, the next question changed. The company was not only asking whether better reporting could be built. It was facing a broader reconciliation problem across thousands of records, along with the uncomfortable possibility that what customers had signed, what people believed, and what the CRM contained were not always the same thing.

The project did not magically resolve the business.

It made the business visible enough that partial understanding could not carry it in the same way.

How It Forms

Forced clarity forms inside systems that have been allowed to operate without a shared picture of reality.

This does not always look careless. In early and growing companies, partial understanding can be a practical adaptation. The business is moving quickly. The product is changing. Customer work is irregular. Processes are being invented. People solve what is in front of them and carry context in their heads because writing it down, integrating it, or reconciling it properly would slow the work.

For a time, that can be rational. The company may not yet need perfect reporting, clean ownership, or fully mature process. A founder can answer the question. A senior operator can remember the exception. A developer can trace the bug. A finance person can work around the inconsistency. Everyone knows the system is not clean, but the mess has not yet become unavoidable.

The danger is that the organisation confuses survivable ambiguity with harmless ambiguity.

The longer the business operates this way, the more work gets built on top of unclear foundations. Customer records drift away from operational reality. Product behaviour depends on assumptions no one has recently tested. Ownership sits between teams. Reports reconcile only because someone knows which numbers to ignore. The organisation functions, but it functions by leaving parts of itself unexamined.

Forced clarity appears when something finally connects those fragments. It may be a proper report, a technical audit, a data migration, a failed demonstration, a board question, a due diligence process, a customer escalation, or a rebuild that exposes how the system actually works.

The trigger can look technical. The consequence is not.

Once the system is visible end to end, the organisation is not dealing with a missing report, a messy spreadsheet, a bad integration, or one unclear owner. It is dealing with what those things had been concealing. The work shifts from finding information to accepting what the information now reveals.

Early Signals

Forced clarity approaches when simple questions start travelling further than expected.

Someone asks who owns a number, why two systems disagree, which process is correct, whether a customer record can be trusted, or what would happen if a particular assumption changed. The question sounds contained, but the answer requires more people, more history, and more reconciliation than expected.

The organisation may also keep finding local explanations for structural problems. A discrepancy is treated as a data issue. A failed handoff is treated as a training issue. A confusing report is treated as a formatting issue. A product limitation is treated as a bug. Each explanation may contain some truth, but none of them is large enough to explain why the same kind of confusion keeps returning.

Meetings may also harden. Earlier conversations may have stayed exploratory: "Let's understand it", "Let's map it", "Let's clean that up." Later conversations require decisions: who owns this, which source of truth wins, what must be corrected, who tells the customer, what does this mean financially, and what do we stop pretending is still fine?

The emotional signal is important too. Visibility feels threatening before anyone has done anything with it. People become careful about what a report might show, what an audit might imply, or what a proper reconciliation might force into the open. The organisation says it wants clarity, but part of the system is bracing for consequence.

What Gets Misread

Forced clarity can look like the solution.

It is easy to believe that once the report exists, the audit is complete, the data is reconciled, or the system is visible, the hard part has been done. In reality, clarity may only be the point at which the hard part cannot be avoided.

Seeing the system is not the same as repairing it. A clean report can expose dirty data. A technical audit can reveal decisions the team is not ready to own. A process map can show that no one actually controls the full process. A rebuild can make the underlying logic visible enough to prove that the problem was never only implementation.

Blame is the next misread. When clarity arrives late, it can feel like indictment. People may experience the report, audit, or question as an attack because it names what the organisation had been living around. That reaction is understandable, but it can also derail the moment. The visibility becomes the problem, rather than the condition that finally makes repair possible.

The final misread is that ambiguity had been neutral. It rarely was. Ambiguity had been doing work. It allowed teams to move without resolving ownership, allowed leaders to defer uncomfortable choices, allowed different people to maintain different versions of reality, and allowed the organisation to avoid deciding which costs were real.

When clarity is forced, that work stops. The company loses the protection that vagueness had been providing.

Hidden Cost

Forced clarity exposes what ambiguity had been helping the organisation contain. Contradictions that once lived as background discomfort become visible responsibilities.

This cost is practical before it is philosophical. Someone has to reconcile records. Someone has to decide which system is authoritative. Someone has to tell a customer that the previous understanding was wrong. Someone has to explain the financial impact. Someone has to choose whether the product is repaired, rebuilt, narrowed, paused, or abandoned.

But the deeper cost is ownership.

Before forced clarity, the organisation can distribute discomfort across many small uncertainties. After forced clarity, uncertainty concentrates. The business may not know what it wants to do yet, but it cannot honestly say it does not know what the issue is. The question changes from "What is happening?" to "Who is willing to own what is now visible?"

Forced clarity can therefore feel less like insight than burden. It removes the comfort of partial understanding. It makes denial harder. It converts hidden contradiction into explicit work.

It can also reveal how much of the organisation's confidence depended on not looking too closely. That does not mean the confidence was fake. Often it was built from real progress, real customers, real effort, and real capability. But progress can coexist with unresolved incoherence. Forced clarity is the point where those two realities have to be held together.

Recognition

Forced clarity begins with an uncomfortable question: what ambiguity is still carrying us?

The discomfort is the point. It asks where the business is relying on unclear ownership, incomplete reporting, founder memory, informal process, partial data, or unresolved assumptions to keep moving. It does not assume those things are always wrong. Early companies need some looseness. The issue is whether the looseness is still serving learning, or whether it is now protecting avoidance.

The next question changes the purpose of visibility: if this became visible end to end, what would it force us to own?

This shifts the purpose of clarity. The goal is not to know more. The goal is to understand what greater visibility would make unavoidable. Would it expose customer impact, financial misstatement, product debt, unclear accountability, operational fragility, or a strategic choice the company has been delaying?

Finally, watch whether visibility is treated as threat or as responsibility.

The organisation's response to forced clarity determines whether reckoning becomes repair or collapse. If the company punishes the person who made the system visible, debates the tone of the finding, or retreats into defensiveness, it may preserve pride while losing the chance to correct. If it can hold the discomfort long enough to own what has been revealed, clarity can become the beginning of maturity.

Forced clarity does not resolve the reckoning by itself. It does not decide whether the company continues, transforms, narrows, or exits. It simply removes the option of not knowing.

The hard part begins after the system becomes legible. At that point, the problem is not hidden. Someone has to decide what to do with what is now known.

Chapter 28
When Deferral Ends

Reckoning does not usually arrive as a single dramatic event.

It appears as a gradual loss of distance between decisions and their consequences. What was once postponed begins returning in compressed form. Conversations shift. Assumptions surface. Systems that once absorbed tension begin reflecting it back.

The organisation has not suddenly changed.

What has changed is its ability to postpone seeing what it has already created.

Across this part, the pattern has been one of consequence becoming harder to avoid. The bill comes due for decisions that once looked temporary, necessary, or distant. The time between reality and consequence compresses until there is less room to recover. Options narrow as earlier trade-offs quietly decide which futures remain available. Then clarity is forced, not as comfort or insight, but as exposure.

By this point, the organisation may still be moving. It may still have customers, plans, meetings, forecasts, roadmaps, and people trying hard to make the next decision work.

This matters.

Reckoning is not the same as collapse.

It is the end of a particular kind of deferral.

The company can no longer rely on the old distance between story and structure, effort and effect, intention and consequence. The patterns have become visible enough that not seeing them now requires a different kind of refusal.

This is where the human difficulty of the book becomes plain. Recognition does not automatically make the next move easier. Sometimes it makes the next move harder because the organisation now has to choose with fewer illusions available. It has to decide what can be repaired, what must be narrowed, what must be named, and what can no longer be carried.

For some organisations, this becomes maturity. They do not escape consequence, but they stop negotiating with it as if it were still theoretical. They reconnect responsibility to reality. They accept smaller, clearer, more honest choices. They rebuild trust by telling the truth sooner than the system used to allow.

For others, the same visibility becomes fragmentation. People defend the story, punish the signal, keep asking for clarity while resisting what clarity reveals, or continue treating each exposed cost as a local problem. In those cases, the reckoning does not end because the organisation has learned. It ends because the system can no longer absorb the delay.

The difference is not effort.

Most people inside these systems have been making effort for a long time.

The difference is whether effort is finally attached to reality.

The final part of the book begins from that bridge.

Once the patterns are visible, the question changes. It is no longer only, "What is happening here?" It becomes, "How do we respond now that we can see it?"

Part VI turns toward that question.

Not with a promise of easy repair, and not with a checklist that pretends every organisation can be rescued from the same distance. The work begins more modestly than that: by learning how to recognise patterns early enough, speak about them clearly enough, and respond before the system has spent all of its room to move.

Because once deferral ends, the possibility of a different response depends on what the organisation is willing to see next.

Part VI

After Recognition

"The difference is not effort. The difference is whether effort is finally attached to reality."

Chapter 29
Acting on Recognition

Once you start seeing these patterns, the natural impulse is to ask what to do about them, and that impulse makes sense.

Recognition rarely arrives while startups are calm. More commonly, it appears when pressure rises, conversations shift, and something that once felt normal begins to look different. At this point in the book, certain dynamics may be harder to ignore. Once patterns become visible, they tend to appear everywhere: in familiar meetings, in confident updates, and in decisions that make sense locally while compounding elsewhere.

That can create a strong urge to act. You may want to name the pattern, correct it, escalate it, or force clarity around a startup that now appears easier to diagnose than it did before. That instinct is understandable, but usually too fast.

One of the mistakes leaders make when they begin recognising patterns is acting before signal interpretation has finished. They move from recognition straight into correction. They try to solve the visible surface without spending enough time understanding how the pattern is being sustained, who is carrying it, and what costs are already being absorbed to keep it in place.

I prefer the language of orientation before intervention for this reason. The first task is not always to fix, but to see more honestly.

That may sound passive, but it is not. It takes discipline to slow down inside a fast-moving environment, especially one that rewards confidence, output, and decisiveness. Clear seeing is not hesitation; it is what makes any later response more intelligent.

Seeing More Honestly

So begin by noticing what your startup is quietly carrying.

- What keeps getting patched?
- Which concerns keep reappearing under slightly different wording?
- What explanations recur so regularly that they now feel like atmosphere rather than argument?
- Where has curiosity been replaced by reassurance?
- Which parts of the product only feel stable when the path is controlled?
- What is the team compensating for with effort?
- Which questions seem to get safer after the meeting instead of during it?

These are not checklist items in the usual sense. They are prompts for attention. Used well, they reopen perception inside systems that have become too fluent at explaining themselves.

If you want a slightly more structured way to work with them, start with the arc the book has been tracing.

First, look for the story:

What is the company currently telling itself about the product, the market, the team, or the architecture? Which parts of that story are still helping, and which parts are now determining what people are allowed to see? If the same explanation keeps absorbing every new signal, you may be inside Narrative Lock-In.

Second, look for the accumulation:

Where does the language of temporary compromise keep appearing? What is always going to be cleaned up later? Which parts of the system feel close, but only if you avoid looking too closely? If the product requires increasing amounts of care just to preserve the sense of forward motion, you may be inside Incremental Compromise.

Third, look for the human load:

Who is absorbing the cost? Where is strain becoming normal? Which people are acting as the system's extra margin? Are founders carrying more tension than they can articulate? Are teams adapting to conditions that should still feel abnormal? If the company is being held together by what people are tolerating, that is not background noise. It is part of the structure.

Fourth, look for the consequence:

What is becoming harder to defer? Which costs now need to be paid in money, time, trust, attention, or repair? Where are options narrowing even though the conversation still sounds open? What has become visible enough that someone now needs to own it? If ambiguity is no longer carrying the system, you may be approaching reckoning.

Testing The Signal

If you are trying to decide whether what you are seeing is real, begin with recurrence.

A pattern is rarely confirmed by one bad meeting, one failed release, one tense conversation, or one alarming message from someone close to the work.

Startups are noisy places. Good teams can have bad weeks. Strong products can have fragile moments. Pressure can temporarily distort behaviour without yet telling you what kind of system you are actually inside.

The stronger test is whether the same signal keeps returning through different parts of the startup.

- Does the reassuring story keep absorbing concerns that should have changed it?
- Do the same compromises keep reappearing under different names?
- Do the same people keep becoming the extra margin?

- Are consequences becoming harder to defer, easier to feel, or more difficult to keep local?
- Does progress keep depending on controlled paths, private rescue work, or a growing tolerance for what was supposed to be temporary?

When the same concern appears in the story, in the accumulation, in the human load, and in the consequences now returning to the room, you are usually no longer looking at an isolated problem, but at a structure showing itself.

This is one reason it helps to test what you are seeing from more than one angle before drawing a conclusion. A single complaint may be personal. A single unstable release may be recoverable. A single awkward meeting may mean very little. But when the narrative remains overconfident, the workarounds keep expanding, the same people are absorbing the cost, and the consequences are becoming harder to postpone, the signal becomes harder to dismiss as noise.

You do not need perfect proof before you take a concern seriously. But it helps to ask a better question than *is this bad enough yet?*

A more useful question is: *is this recurring in a way that suggests the system is producing it?*

Recurrence strengthens the signal, but it does not automatically make your first explanation correct. Once a pattern seems to fit, ask what else could plausibly produce the same signs, and what evidence would make you less confident in the interpretation you have formed.

The point is not to explain a real concern away. It is to keep the pattern itself from becoming another story through which every new signal is interpreted.

Mistaking Motion For Learning

There is another risk here as well. Once patterns become visible, startups can feel pressure to respond with something immediate and legible. The room wants movement. Leadership wants a path. The wider startup wants reassurance that what has been noticed will now be handled.

That response usually takes one of two forms. The first is continuation. More money goes in. More bugs are patched. More effort is asked of the same people. The company tells itself that one more release, one more stabilisation push, one more round of funding, or one more attempt to reach MVP will justify what has been spent. Sometimes that is perseverance. Sometimes it is sunk cost dressed up as resolve.

The second response carries a very different tone, but not always a different logic. The company decides to fail fast. It restarts, rebuilds, pivots, or changes stack. Sometimes that is exactly the right decision. But not every restart reflects real learning. Sometimes a reset is a cleaner way to avoid the more difficult question of why the same conditions kept producing the same result in the first place.

In some startups, that becomes a kind of reset addiction: the company keeps treating a new beginning as proof of learning, even while the conditions that produced the original result remain largely unchanged.

Continuation and restart should not be treated as opposites by default. One keeps carrying the visible system. The other throws it away. But both can preserve the same decision environment underneath: the same pressure to protect momentum, the same weak translation between vision and buildability, the same bypassing of discipline, the same reluctance to absorb unwelcome signal early, and the same habit of treating urgency as proof that even more speed is required.

Seen this way, failing fast is not the same as learning fast. And continuing is not the same as commitment.

The more useful question is not whether to keep going or start again. It is whether the startup has become more honest about what produced the condition it is now responding to. If the answer is no, then continuation may deepen the problem and restart may only repeat it.

Visible action can calm a room long before it improves a system. The same mistake can take other forms.

Once a company feels the pressure to respond, many other moves can also borrow the appearance of intervention without absorbing the pattern itself.

- A new reporting layer can be added while the same filtered signals continue travelling upward.
- A new leader can be hired while the same decision environment makes honest escalation expensive.
- A task force can be formed while no one is willing to narrow the promise that created the strain.
- A process can be announced while exceptions continue being made for the same urgent personalities and symbolic deadlines.
- An audit can be commissioned and treated as action, even if the organisation has no appetite to integrate what the audit reveals.

Any one of these moves may be useful. The risk appears when the organisation starts treating the visible response as proof that the underlying condition is being addressed.

At that point, intervention becomes theatrical. The company can point to motion, effort, seriousness, and new structure. What it may be unable to do is alter the assumptions, incentives, and translation gaps that allowed the pattern to form in the first place.

For that reason, the practical question is not only *what are we doing about it?* It is also *what is actually changing that would make the pattern harder to reproduce?*

Making The Signal Discussable

Seeing a pattern is one challenge; talking about it well is another.

One reason people stay silent after recognition is that they assume the only way to raise a serious concern is to make a dramatic claim. They think they need to arrive with a conclusion strong enough to justify the discomfort they are about to create.

Usually, they do not need that. In many startups, the most useful language is quieter than that. It is better to name what you are observing before you name what you think it means.

- The demos only seem to hold if the path is controlled.
- We keep describing this as temporary, but the same issue has returned three times under different names.
- The team appears to be compensating for something structurally, not just pushing through a busy period.

That kind of language keeps the conversation closer to shared evidence and farther away from identity threat. It is easier for a room to hear *this pattern keeps recurring* than *this company is failing.* It is easier to ask *what are we relying on here that no longer feels temporary?* than to declare *this approach is broken.*

This is not timidity; it is precision strong enough for the signal to survive contact with the room.

Questions are more useful than accusations.

- What keeps needing rescue here?
- What are we treating as temporary that has started behaving like structure?
- What is becoming harder to say directly in the meeting than outside it?
- What are people closest to the work compensating for that the wider system may not yet fully see?

Used well, questions like these do not weaken the concern. They make it harder to dismiss.

They also help separate interpretation from performance. The goal is not to sound insightful, brave, or prophetic. The goal is to make the signal discussable before the system is forced to confront it more brutally.

Where Leverage Lives

Once the signal becomes discussable, the next question is where intervention actually lives.

People can imagine intervention too dramatically. They imagine that once a pattern is visible, someone must either rescue the company, overturn the strategy, confront the founder, or force a complete change of direction.

Sometimes that scale of change is necessary. In many cases, though, the useful intervention happens earlier and at a smaller unit than that.

- It may be a better question asked at the right moment.
- It may be slowing a commitment that was about to become irreversible.
- It may be separating a demo from a product claim before the two harden into the same thing.
- It may be narrowing a promise, exposing a hidden dependency, insisting that a recurring workaround be named as structural rather than temporary, or refusing to let a fragile condition keep borrowing the language of momentum.

In other words, intervention does not always begin with control; it begins with precision.

The system is usually not changed all at once. It changes when someone makes it harder for the next distortion to pass through unchallenged in its usual form.

Here, leverage is more useful than force. Force tries to overpower the whole pattern at once. Leverage interrupts the point where the pattern keeps converting ambiguity into permission.

- Where is the commitment being made too quickly?
- Where is confidence outrunning evidence?
- Where is a workaround being preserved by language that no longer fits reality?
- Where is a room about to move forward on an assumption that no one has properly tested?

Those points may be smaller than they first appear. They are also where intervention tends to become most real.

Reading The Room

If you are trying to judge the condition of a room in real time, do not listen only for what is being said. Look for how the room is carrying what is being said.

In healthy moments, concerns can usually be voiced directly enough that they remain recognisable when the meeting ends. Questions sharpen discussion. Disagreement may be uncomfortable, but it has enough legitimacy to stay in the room.

When strain is becoming structural, the signals can shift in subtler ways.

- People closest to the work start softening their language.
- Concerns are raised as passing comments rather than full arguments.
- Reassurance arrives faster than curiosity.
- The hardest questions appear after the meeting rather than during it.

- Specific technical concerns get translated upward into safer, thinner summaries.
- The same one or two people keep carrying the clarifications, caveats, or rescue context that allows the conversation to keep moving.

Each signal is partial on its own. Taken together, though, they can tell you that the room is no longer processing strain cleanly.

If you hold authority, silence is not neutral data. Power changes what reaches you. People learn quickly which concerns travel cleanly, which ones return softened, and which ones quietly increase their social cost. Leaders who want the truth have to ask for the harder version of it and make it usable when it arrives, or the room will keep feeding them safer summaries.

This is also why tone can be so informative. A meeting does not need to become openly conflictual to reveal that something is wrong. Sometimes the clearest signal is the opposite: caution without clarity, agreement without conviction, movement without relief.

When a room keeps moving but seems to get thinner each time it does, pay attention.

A system can feel this way when it is becoming more dependent on interpretation, buffering, and private correction than on shared understanding.

Without Authority

All of this remains true even when you are not the founder, the executive, or the person with formal authority to change the trajectory directly.

In fact, many of the people who recognise these patterns earliest are not the ones with the power to decide.

They are senior engineers, delivery leads, product managers, architects, or trusted operators sitting close enough to the work to feel the distortion forming before it becomes legible at altitude.

That position can be deeply frustrating. It is easy to assume that if you cannot make the final call, then recognition has little use beyond private correctness. But even without authority, you may be able to do important work.

- You may be able to preserve signal that would otherwise be softened into something safer.
- You may be able to ask the question that slows an overconfident commitment before it hardens.
- You may be able to make a hidden dependency visible to people who would otherwise continue past it.
- You may be able to stop carrying a false sense of personal responsibility for conditions you did not create and cannot individually repair.

The work may be smaller than rescue, but more important than silence. It may be writing the dependency down in a shared place instead of keeping it in private memory. It may be asking whether a feature is core to version one, or whether it only makes the story more impressive. It may be separating a demo path from a product claim before those two things become impossible to untangle. It may be saying, calmly and early, that a technology choice is slowing learning rather than accelerating it.

None of that requires final authority. It does require enough discipline not to turn private recognition into private compensation.

What you cannot do is carry the whole burden of correction on behalf of the system. That distinction matters.

People without final authority can slowly become the startup's private truth-tellers, rescue layer, or emotional shock absorbers. They see the strain, compensate for it, explain around it, and quietly keep the system functional enough that others can continue misreading it.

That may feel responsible in the short term. Over time, it can also make the pattern harder to surface.

Sometimes the most useful act is not heroic correction.

- It is refusing to make the signal disappear.
- It is naming the condition accurately.
- It is documenting the dependency cleanly.
- It is asking the question in the room rather than only after it.
- It is being careful not to turn your private recognition into one more invisible subsidy the system can keep consuming.

Sometimes that also means accepting the limits of your position sooner than you want to. If a system keeps asking you to absorb what it refuses to recognise, the healthier act may not be another rescue attempt. It may be making the cost visible, setting a boundary, or eventually stepping away before the pattern consumes more of you than the company is willing to see.

Better Outcomes

One final distinction is worth making here. When people see these patterns, they may imagine that a good outcome must look like rescue.

The product stabilises. The team regains energy. The original promise survives intact. The company proves it really was the exception after all. Sometimes that does happen, but it is not the only form a better outcome can take. In many cases, the healthier outcome is more modest and more honest than that.

- A startup narrows its scope before the strain becomes unrecoverable.
- A founder stops defending a story the system can no longer support.
- A team separates what is demoable from what is durable.

- A product is narrowed to something the startup can actually carry.
- A startup accepts restructuring, acquisition, or a smaller identity instead of preserving the old ambition at any cost.
- A project ends earlier, more cleanly, and with less avoidable damage than if everyone had continued pretending.

These outcomes rarely feel heroic in the usual startup sense. Some of them may feel like loss. Some may look, from the outside, like retreat.

But not every good outcome is expansion. Not every honest response is acceleration. And not every meaningful intervention ends with the original story intact.

Sometimes the better outcome is not that the startup becomes everything it once said it would be. It is that reality is allowed to set the terms sooner, while choice, dignity, and some room to act remain.

I have seen this happen in quieter ways than the usual startup story knows how to celebrate.

In one fast-growing business, a reporting rebuild was expected to provide better visibility over a messy operating system. It did that, but not in the way people first expected. Once the data could be traced end to end, the company discovered that the problem was not only reporting. The inputs themselves were wrong. Customer records, pricing information, cost assumptions, and manual handoffs had drifted during growth until the business could no longer trust the system it had been using to understand itself.

That was not a heroic moment. It was uncomfortable. The new visibility created more work, not less. But it also gave the organisation something it had not had before: a shared picture of the whole process. Once the full path was visible, repair could begin in a more honest place. The company could stop treating the issue as a local reporting problem and start aligning the inputs that produced the result.

Better can look like that.

Another version happens at a smaller scale. A team may spend months on a feature that sounds impressive, only to realise it is not central to the product the company actually needs to prove. Stopping that work can feel like loss because effort has already been spent. But if the decision returns attention to the thing the product must become first, it is not failure. It is scope becoming honest again.

The same can be true of technical choices. Sometimes the better move is not the more sophisticated architecture, the more serious platform, or the tooling that makes the company feel closer to its future scale. Sometimes the better move is choosing the simpler path the team can understand, change, and learn from quickly. Not because simple is always better, but because the company has not yet earned the complexity it is tempted to wear.

From there, one thing becomes clearer: clarity creates leverage. Not certainty, and not a guarantee, but leverage.

If you can see what the system is really doing, you have a better chance of asking the question that slows the room down at the right moment. You have a better chance of naming the assumption underneath the decision instead of arguing only about the surface choice. You have a better chance of recognising when a product is being made to look ready rather than being made ready.

The empowering part of this work is not immunity to the patterns, and not perfect foresight, but a better chance of not mistaking visibility for negativity, caution for weakness, or discomfort for irrelevance.

If you want a simple next step after reading this book, try carrying one or two questions into your next relevant meeting.

- What are we assuming here that no longer deserves to remain implicit?
- What is being held together by effort rather than by structure?
- What are people closest to the work carrying that the wider system is not yet fully seeing?
- What consequence are we treating as avoidable?

You do not need to ask them dramatically. In fact, it is usually better if you do not. The goal is not to perform insight. The goal is to create a little more room for honest signal.

The weather first becomes discussable there.

What follows is a final reflection on what these patterns ask of us once they become visible.

Epilogue

"Not certainty. Not immunity. Just a little more honesty, a little earlier, where it still has time to matter."

By the time a startup reaches its visible breaking point, most of the important work has already been done.

Not the work the startup hoped to be known for, but the quieter work of pattern formation.

The story has been told. The assumptions have been defended. The trade-offs have been normalised. The burden has been transferred into code, into process, into meetings, and eventually into people. What looks sudden from a distance has usually been taking shape for a long time.

This book has tried to make that argument.

But if the book stopped there, it would risk making these dynamics feel more contained than they are.

The core stories in this book come from ventures and delivery environments I lived inside. The later conversations widened the field of recognition. They showed the same dynamics appearing in startups with large ambition and fragile foundations, internal innovation teams inside larger organisations, service environments, agencies, and companies that never fully became what they said they were becoming. They also surfaced founders working on meaningful products without the language or visibility to inspect the systems they had paid for. The settings varied. The mechanics did not.

Again and again, the same deeper structure appeared.

A simpler path existed, but lost to the more impressive one. A prototype, a roadmap, or a polished layer created confidence that the underlying system had not earned. Design, product, and engineering became translation layers for a vision that still had not been made buildable. Credibility was borrowed from brands, past success, prestigious hires, or external partners. The appearance of progress remained easier to generate than the conditions that make progress real.

And when those conditions weakened, the cost rarely stayed where the original decisions had been made. It moved downward into delivery teams, sideways into product and design, outward into customers, and inward into the private mental load of the people trying to keep the system coherent.

It moved into founders too, though in a more difficult form. Again and again, one truth became worth naming directly: the person most responsible for the trajectory may also be one of the people most trapped inside it. Many founders are not standing outside these patterns. They are carrying belief from within them, while trying to convert uncertainty into direction for everyone else.

Responsibility remains, but this changes how we understand the problem.

The Founders' Folly is not, in the end, a book about foolish people. It is a book about what happens when confidence, pressure, ambiguity, and power combine in ways that make honest signal harder to keep alive. It is about what happens when a startup becomes more practised at preserving momentum than at testing whether the thing being carried forward still deserves to be carried in the same way.

For that reason, so many of these stories resist the simple categories people prefer.

Not every startup collapses cleanly. Some survive into the world, but in a scarred and expensive halfway state. Some products work well enough in the field and still fail because the wider startup cannot hold, sell, or sustain them. Some teams do strong work inside systems that are already too fragmented to convert that work into durable progress. Some reckonings end in shutdown. Others end in reduction, bounded continuation, or a quieter admission that the startup is smaller than the story it once told about itself.

Even that belongs inside the book's argument. It is part of what the book has been trying to reveal.

The reckoning is not always the end of the startup — it is the end of a particular distance between belief and consequence.

This is why the book has preferred orientation before intervention.

Once you see these patterns, the goal is not to become cynical, omniscient, or permanently suspicious. It is not to drain ambition out of startup work until only caution remains. It is to become less willing to confuse motion with health, polish with readiness, hiring with clarity, or confidence with proof. It is to notice sooner when the organisation is being held together by translation gaps, workarounds, silence, or loyalty that has become too expensive.

It is also to notice what happens to people when those conditions persist.

One of the strongest themes across the wider field was not only burnout inside a single startup, but carryover. Disappointment compounds. Redundancies reshape how people enter the next role. Repeated misalignment changes how fully they trust, contribute, or push back the next time. Human debt does not always end when somebody leaves. Sometimes it becomes part of what they bring with them afterwards.

That matters because startups do not only produce products, valuations, and outcomes. They also produce ways of working, thresholds of honesty, and inherited expectations about what people should quietly absorb.

Across the people and projects behind this book, and in my own life, another truth kept surfacing alongside the structural one. Many of the people carrying these startups looked successful on paper. They were capable, analytical, respected, productive, and trusted with significant responsibility. But the strain was accumulating in them too: in sleep, mood, relationships, marriages, family life, and the private effort required to stay coherent while continuing to perform.

The pattern appeared frequently enough to be difficult to ignore, especially among men whose lives had been organised around competence, output, and the pursuit of success. They had learned to convert pressure into performance for so long that the internal cost could remain hidden even from themselves.

When the startup narrowed, stalled, failed, or simply scarred its way forward, some of the real reckoning was waiting elsewhere: in exhaustion, in addiction, in emotional withdrawal, in the slow work of rebuilding a life that had been shaped too long around strain.

I do not read that only as someone who observed it in others. I read it as someone who eventually had to reckon with it personally. The cost of carrying pressure, ambition, and analytical control too far does not always announce itself while the work is still being praised. Sometimes it becomes visible only later, when functioning starts to collapse, when the inner life has gone thin, and when recovery takes far longer than the old success-driven self would ever have allowed.

This is part of what I want the book to name. These patterns do not only deform products, teams, and startups. They can also deform the people who become too practised at carrying them. And if there is any deeper use in seeing them earlier, it is not only that a startup may make better decisions. It is that a person may not have to spend years rebuilding what the pursuit of success quietly wore away.

That human cost is part of the same argument, not separate from it.

If this book has done its work well, it has not merely named a set of patterns in struggling startups. It has made certain forms of strain harder to dismiss as isolated, personal, or inevitable. It has made it easier to see that what looks like bad luck at the end was structured much earlier through choices that felt local, justified, and temporary at the time.

Recognition does not give anyone full control. It does something more modest, and perhaps more useful: it returns consequence to the room earlier.

It makes it harder for a polished demo to settle the question by itself. Harder for a prestigious name to stand in for proof. Harder for the system's extra margin to remain invisible once you realise it is human. Harder for the weather to be misread simply because the dashboard still looks calm.

It may not sound dramatic enough for an ending, but it is enough.

Sometimes enough means a startup narrows before it overcommits further. Sometimes it means a founder hears a technical concern without forcing it to become optimism on the way up. Sometimes it means a team stops mistaking repeated rescue work for normal execution. Sometimes it means the organisation admits what it is actually capable of carrying, and what it is not. Sometimes it means ending something earlier, more cleanly, and with less avoidable harm.

Not certainty. Not immunity. Just a little more honesty, a little earlier, where it still has time to matter.

The weather will never stop changing.

But the longer I have spent around startups, teams, and the stories they tell about themselves, the less I think the deepest problem is volatility on its own.

It is the habit of calling the weather fine long after the pressure has changed.

If this book helps make that harder to do, it has served its purpose.

Appendix

Glossary

This glossary defines key terms as they are used inside *The Founders' Folly*. It is intended as a quick recall tool while reading, not as a full index of every idea in the book.

Core Terms

The Founders' Folly The Founders' Folly is seeing too late what was forming early. It is the delayed recognition of patterns that capable founders, leaders, and teams could have noticed while there was still more room to act.

Founder In the narrow sense, the person or people who started the company. In this book, the term also reaches toward people with real influence over a startup's direction: executives, investors, senior technical leaders, and others whose decisions materially shape the company's trajectory.

Folly Not stupidity or personal incompetence. Folly refers to predictable patterns of interpretation, pressure, confidence, compromise, and consequence that can make reasonable decisions accumulate into damaging outcomes.

Pattern A recurring structural dynamic that appears across startups. A pattern is not one mistake or one bad decision; it is a shape of behaviour, interpretation, and consequence that repeats under pressure.

Keystone Pattern One of the book's primary structural patterns. Keystone patterns help explain why different behaviours, symptoms, and local variations can share the same larger shape.

Signal An observable clue that a pattern may be forming. Signals usually appear before damage is obvious.

Early Signal A signal visible while choice still exists. Early signals are often small: repeated language, softened concerns, recurring workarounds, hidden compensation, or a question that keeps returning without changing the plan.

Recognition The act of seeing a pattern clearly enough to name it, test it, or make it discussable before consequence fully arrives.

Weather The book's central metaphor for the shifting conditions around a startup: pressure, confidence, strain, momentum, risk, and the signals that show something is changing before the dashboard reflects it.

Dashboard The visible instruments a startup uses to understand itself: metrics, runway, status reports, demos, roadmaps, and updates. The dashboard can look calm while the underlying weather is already changing.

Structure The operating reality beneath visible activity: incentives, processes, architecture, ownership, decision paths, dependencies, and the way work actually moves.

Strain Pressure that has started to affect the system or the people inside it. Strain may be technical, organisational, emotional, financial, or strategic.

Reading The System

Altitude The distance between a person's role and the daily reality of the work. Higher altitude usually means broader visibility but less direct contact with the friction inside the system.

Altitude Distortion The distortion that appears when the same reality looks different depending on how close someone is to the work. What feels like flexibility at leadership altitude may feel like instability to the team carrying the implementation.

Leadership Altitude The perspective of founders, executives, board members, investors, or senior leaders who see summaries, outcomes, and strategic movement more clearly than local operating friction.

Close To The Work The position of people who feel the product, architecture, process, customer reality, or team strain directly. They often see weak signals before those signals become visible at higher altitude.

Signal Interpretation The work of deciding what a signal means. The same signal can be dismissed as noise, explained away as temporary, or recognised as evidence of a pattern.

Orientation Before Intervention The discipline of understanding what pattern is forming before rushing to fix the visible surface. Orientation asks what is being sustained, who is carrying it, and what costs are already being absorbed.

Intervention An action that makes a pattern harder to reproduce. Intervention does not always mean rescue or confrontation; it may be a better question, a slower commitment, clearer ownership, or a signal kept alive.

Leverage The point where a small, precise action can interrupt a larger pattern. Leverage is more useful than force when the pattern is still forming.

Without Authority The position of seeing a pattern without having final power to change the company's direction. The book treats this as difficult but not useless: people without authority can still preserve signal, name conditions, and avoid becoming invisible compensation for the system.

Better Outcome Not necessarily rescue, scale, or a triumphant turnaround. A better outcome may be a narrower scope, cleaner ending, earlier boundary, more honest product shape, or less avoidable human cost.

The Pattern Arc

Pattern Arc The book's organising movement from early interpretation to accumulated cost, human impact, and eventual reckoning.

Allusion / Allusions A signal, story, image, demo, or fragment of progress that points the mind toward a larger story. In the book's first pattern stage, allusions shape what the startup sees, trusts, and explains away.

Illusion The moment a larger story is mistaken for reality. The book uses this distinction deliberately in The Inevitability Illusion: an allusion points toward a story; an illusion mistakes the story for proof.

Accumulation The stage where compromises, workarounds, incentives, and process exceptions begin to compound into structural fragility.

Human Cost The stage where accumulated strain moves into people: their attention, patience, memory, judgement, recovery, trust, and willingness to speak plainly.

Reckoning The stage where accumulated cost can no longer be deferred or distributed quietly. Reckoning is not always collapse; it is the point where consequence becomes unavoidable enough to require ownership.

After Recognition The final movement of the book, focused on what becomes possible once patterns are visible: better questions, cleaner signals, more honest choices, and sometimes less damaging outcomes.

Keystone Patterns

Narrative Lock-In The point where a startup's story stops being a live explanation and becomes the frame through which new information is interpreted.

The Inevitability Illusion The distortion that forms when early progress makes success feel more settled than it has actually become.

Borrowed Confidence Confidence drawn from people, brands, vendors, tools, past success, or external credibility rather than from what the current system has actually proven.

Incremental Compromise The pattern where small, reasonable trade-offs compound into a system that is harder to change, trust, or recover.

Process Erosion The gradual weakening of the processes that are supposed to preserve visibility, ownership, review, and shared understanding.

Incentive Drift The pattern where energy follows what is rewarded or admired, even when that work no longer points toward the outcome the company claims to want.

Human Debt Accumulation The transfer of unresolved structural cost into people, who compensate through attention, memory, patience, explanation, care, and endurance.

Normalised Overextension The pattern where exceptional stretch becomes the assumed baseline of work.

Emotional Withdrawal The point where people remain present and capable, but stop bringing the full truth of what they know into the formal life of the organisation.

The Bill Comes Due The moment when costs that have been distributed, deferred, or explained as temporary become concrete enough that someone has to pay attention to them.

Compression of Consequence The pattern where the delay between reality and consequence shortens. What once surfaced slowly now lands quickly, narrowing the mental and organisational space available to respond.

The Narrowing of Options The pattern where the company still speaks as if many choices remain open, while earlier decisions have already reduced the meaningful field of action.

Forced Clarity The moment when ambiguity stops carrying the system and the organisation can no longer avoid seeing what must now be owned.

Supporting Terms

Technical Debt The cost of earlier technical shortcuts, complexity, or deferred cleanup. The book often uses technical debt as a contrast to human debt.

Human Debt The human version of accumulated debt: unresolved structural burden stored in people rather than in code, process, or planning.

Human Margin The spare attention, patience, judgement, recovery, and emotional capacity people need in order to think clearly and respond well under pressure.

Invisible Support Work The hidden explanation, checking, translation, repair, reassurance, and follow-up required to make visible work appear to function.

Private Compensation The quiet work people do to keep a system functioning without forcing the system to recognise what they are absorbing.

Structural Cost The cost created by the shape of the system itself: dependencies, incentives, process gaps, architectural choices, unclear ownership, or accumulated decisions.

False Surprise The shock experienced by people who encounter a consequence late, even though people closer to the work have been living with the underlying reality for some time.

Reset Addiction The habit of treating a restart, rebuild, pivot, or stack change as proof of learning without changing the conditions that produced the original problem.

Failing Fast / Learning Fast Failing fast means stopping or restarting quickly. Learning fast means understanding what produced the condition in the first place. The book treats these as related but not equivalent.

Structural Acceleration The compression of startup dynamics when tools, AI, capital, or market pressure allow a company to move faster than its systems of judgement, ownership, and learning can safely absorb.

Pattern Memory Jogger

This section is for returning to the book after the first read.

Start with the signal, not the label. If several patterns seem relevant, begin with the earliest one the company can still influence. A visible reckoning may point back to accumulated compromise. Human cost may point back to incentive drift or process erosion. Borrowed confidence may be supporting an inevitability illusion. The name matters less than the movement it reveals.

It is not a substitute for the chapters, and it is not a diagnostic test. If you are trying to work out what a live situation may be showing you, use the Pattern Diagnostic Index. This section has a simpler job: to help you remember the shape of the book quickly.

The Pattern Arc

The core arc of *The Founders' Folly* moves through four stages:

Allusions — Accumulation — Human Cost — Reckoning

That arc is not a strict sequence. Patterns overlap. A company may be carrying human cost before the reckoning is visible to leadership. Accumulation may be happening in one part of the business while another part is still caught inside the language of allusions.

The arc is a way to read movement, not a checklist to follow in order. It shows how a startup can drift from interpretation, through accumulated structure and human cost, toward consequence that can no longer be kept at a distance.

Allusions

Allusions are the early stories, symbols, demos, and fragments of progress that shape how a company interprets what it sees.

At this stage, the future may feel visible before the present system has proved it. A demo, a founder story, a market signal, a credible person, or an impressive partner can point the company toward a larger meaning. The danger is not belief itself. The danger is belief becoming a filter before the evidence has earned that role.

If allusions harden, the company may stop learning from reality and start explaining reality until the story remains intact.

Accumulation

Accumulation begins when reasonable trade-offs, exceptions, incentives, and workarounds start compounding into structure.

Nothing has to look foolish in isolation. A shortcut may be sensible. A process exception may be necessary. A feature may deserve priority. The signal is repetition: the same kind of compromise returning, carrying more weight each time, until later work depends on what was meant to be temporary.

If accumulation continues, the system becomes harder to change, trust, explain, or reverse.

Human Cost

Human cost appears when unresolved structural strain moves into people.

The company may still appear to be functioning. Work ships. Meetings happen. Customers are served. But continuity increasingly depends on people absorbing what the system does not hold: memory, judgement, patience, emotional steadiness, translation, rescue work, and private recovery.

If human cost continues, human margin becomes the hidden operating model.

Reckoning

Reckoning arrives when deferred cost becomes concrete enough to require ownership.

The bill may appear as money, time, customer exposure, technical rebuild, organisational disruption, lost trust, or a decision that can no longer be delayed. The company may still discuss many options, but fewer of them remain genuinely available.

If reckoning continues without honest ownership, consequence compresses, ambiguity fails, and clarity is forced.

Around the Arc

Two parts of the book sit around the core arc.

Reading the Weather comes before the arc. It teaches the reader how to see signals, altitude distortion, and structure beneath visible progress.

Before a pattern can be understood, the organisation has to notice that something in the conditions has changed. Reading the Weather is about that first act of perception: looking past dashboards, summaries, optimism, and local explanations toward the pressure, strain, and distortion already forming around the work.

If this part is missed, the rest of the arc can look too sudden. The story appears to harden all at once. Compromise appears to accumulate from nowhere. Reckoning appears as surprise. Reading the Weather is what helps the reader notice that the conditions were already speaking before the pattern had a name.

After Recognition comes after the arc. It asks what becomes possible once patterns are visible, and how recognition can change the next decision rather than merely describe the last one.

Recognition does not automatically fix a startup. It can arrive late, unevenly, or without enough authority to change the whole system. But it can still alter the quality of the next question. It can make a signal discussable, slow a commitment, preserve evidence, name a dependency, or prevent a reset from being mistaken for learning.

If this part is missed, the book can sound only diagnostic. After Recognition is where the book turns back toward agency without promising rescue. The point is not to escape every pattern. It is to see clearly enough that the next decision is made with more honesty than the last one.

Keystone Patterns

Use these notes when you want to remember how the patterns fit inside the arc.

Narrative Lock-In

Narrative Lock-In is the pattern where the company's story stops learning.

Early on, the story helps people make sense of uncertainty. It gives the team direction, confidence, and a reason to keep going. The problem begins when every new signal is interpreted through the story instead of being allowed to change it.

The Inevitability Illusion

The Inevitability Illusion is the pattern where progress makes the future feel more settled than it is.

It appears when the company has something visible: a working demo, an impressive prototype, customer enthusiasm, a strong roadmap, or enough momentum to make the next stage feel obvious. The future starts to feel like it is arriving, even though the product, market, system, or customer path has not yet proved itself under real conditions.

Borrowed Confidence

Borrowed Confidence appears when credibility from elsewhere substitutes for evidence inside the current system.

The confidence may come from an experienced founder, respected hire, well-known vendor, impressive investor, serious consultancy, powerful platform, AI tool, or past success. All of those things can matter. None of them removes the company's responsibility to understand its own dependencies, constraints, and operating reality.

Incremental Compromise

Incremental Compromise is the pattern where small, reasonable trade-offs become the structure later work depends on.

It usually does not begin with a reckless decision. It begins with pressure. Ship this release. Support this customer. Make this demo work. Delay the cleanup. Carry the manual step for now. Each decision can be defensible on its own. The pattern forms when those decisions are not revisited and the next layer of work starts assuming them.

Process Erosion

Process Erosion appears when the official process remains visible but no longer governs how work really moves.

The meeting still exists. The ticket still exists. The review still exists. But urgent work starts travelling through side channels, direct requests, private messages, founder preference, or after-hours decisions. People learn where authority actually sits, and the formal process becomes something updated after reality has moved.

Incentive Drift

Incentive Drift is the pattern where energy follows what is rewarded rather than what most strengthens the company.

The work that gets attention may be impressive, fundable, visible, or easy to explain: the new feature, the investor demo, the partnership, the AI layer, the metric that looks good in a meeting. Meanwhile, quiet work such as reliability, supportability, documentation, testing, onboarding, cleanup, or operational clarity keeps losing because it is harder to admire.

Human Debt Accumulation

Human Debt Accumulation appears when unresolved structural cost is stored in people.

The team may look staffed. The plan may look covered. The product may keep moving. But the system only works because certain people keep lending extra memory, judgement, patience, translation, reassurance, review, or emotional steadiness. Their capability hides the gap.

Normalised Overextension

Normalised Overextension is the pattern where exceptional effort becomes the assumed baseline.

It begins with a push that feels necessary. A late night. A weekend. A compressed release. A difficult customer save. A founder stretch. A team that finds a way. The danger comes later, when the next plan quietly assumes the same level of stretch without naming it as exceptional or building in recovery.

Emotional Withdrawal

Emotional Withdrawal appears when people remain present and capable but stop bringing the full truth into the formal life of the organisation.

The company may become calmer. Meetings may become smoother. Disagreement may reduce. But the strongest version of reality may have moved elsewhere: private messages, side conversations, one-to-ones, technical comments, or conversations after the meeting.

The Bill Comes Due

The Bill Comes Due is the moment when distributed cost becomes concrete enough to demand payment.

For a long time, the cost may have been scattered across small delays, workarounds, technical friction, customer support, team strain, missed cleanup, and quiet explanation. Then it condenses into something visible: a rebuild, a funding need, a delivery failure, a support burden, a customer issue, or a strategic choice that cannot be postponed.

Compression of Consequence

Compression of Consequence is the pattern where the delay between reality and consequence shortens.

Earlier in the life of a problem, the gap may be forgiving. A flaw remains internal. A workaround buys time. A team absorbs the friction. Later, the same kind of issue reaches customers, leaders, revenue, trust, or adjacent teams much faster. The system has less room to explain, soften, or absorb what is happening.

The Narrowing of Options

The Narrowing of Options appears when choices remain nameable after they have stopped being realistically available.

A company may still talk as if many paths remain open: rebuild later, pivot later, simplify later, hire later, fix the architecture later, return to the customer problem later. But earlier decisions may have made those options too expensive, disruptive, politically difficult, or dependent on energy the company no longer has.

Forced Clarity

Forced Clarity arrives when ambiguity can no longer carry the system.

For a while, unclear ownership, partial reporting, founder memory, informal process, incomplete data, or unresolved assumptions may allow the company to keep moving. Then something forces the whole picture into view: an audit, report, customer escalation, technical review, funding process, leadership change, or operational failure.

Pattern Diagnostic Index

This index is for the moment when you do not have time to reread the whole book, but something about the work has started to feel familiar.

You may be inside a difficult meeting, reviewing a project that no longer makes sense, dealing with another failed release, or trying to understand why a capable team has become quiet, brittle, or exhausted. Start with the phrase you keep hearing, the behaviour you keep seeing, or the condition that feels familiar. Then use the chapter pointers to find the fuller pattern underneath it.

This is not a diagnostic test, and one signal does not prove a pattern. Look for recurrence. When the same signal appears across language, decisions, process, technical reality, and human behaviour, it is less likely to be an isolated problem.

Treat the pattern you recognise as an explanation to test, not a conclusion to defend. Ask what else could plausibly explain the same signals, and what evidence would make you less confident that this is the pattern you are seeing.

The Allusions

Narrative Lock-In

You might hear:

- "We are different."
- "That happened to other startups, but it will not happen here."
- "That is not really our situation."
- "The market just does not understand what we are building yet."
- "Customers will understand it once they see the full version."

- "The concern makes sense, but it does not change the bigger story."

You might see:

- The same concern returns, but the explanation never changes.
- New information is treated as a variation of the existing story.
- Customer hesitation, technical friction, and delivery strain are all explained through the same frame.
- People learn which evidence is allowed to change the plan and which evidence is only allowed to be explained.
- The moment that should rename the problem passes without being named.

Ask:

- Does the story still change when reality teaches the company something?
- What evidence would force us to revise the story, not just explain it?
- Which concerns keep returning while the answer stays the same?
- What has changed in the product, market, customer, or system while the old explanation still governs the work?
- Are we using the story to interpret reality, or to protect ourselves from it?

For the fuller pattern, see **Chapter 9 — Narrative Lock-In**.

The Inevitability Illusion

You might hear:

- "We are close."
- "One more push and we will have the MVP."
- "Once this next piece lands, everything will come together."
- "The platform is starting to take shape."
- "We are building for where the company is going, not where it is now."
- "We just need the product to catch up with the vision."

You might see:

- The demo works, but only on a controlled path.
- The roadmap becomes more persuasive than the product.
- Foundational questions keep being treated as sequencing issues.
- Infrastructure, hiring, tooling, or marketing starts implying maturity before version one is clear.
- The future becomes emotionally convincing before the current product has earned confidence.

Ask:

- What has this progress actually proved?
- What is version one?
- Who is the first customer?
- What problem are we solving first?
- What has been proven outside the demo path?
- Has the first customer path become repeatable, or does every new case still require custom effort and special interpretation?

For the fuller pattern, see **Chapter 10 — The Inevitability Illusion.**

Borrowed Confidence

You might hear:

- "They have done this before."
- "The vendor will sort it out."
- "The new hire knows how to scale this."
- "The AI tool makes this much easier now."
- "The agency knows what they are doing."
- "We have serious people around the table now."

You might see:

- A respected person, vendor, investor, platform, or tool reduces scrutiny rather than improving clarity.
- The company talks more about who is involved than what has been proven.
- External credibility makes present risk feel safer than it is.
- A partner's capability is treated as if it has already transferred into the company.
- The serious-looking wrapper dulls the need to inspect the current system.

Ask:

- What confidence have we earned here, in this system, under these conditions?
- Is external credibility improving diagnosis, or reducing scrutiny?
- Which known constraints feel safer because a credible person, vendor, platform, tool, or partner is attached?
- Are we inspecting the dependency, or trusting the reputation around it?

For the fuller pattern, see **Chapter 11 — Borrowed Confidence.**

The Accumulation

Incremental Compromise

You might hear:

- "It is only temporary."
- "We will clean it up after the next release."
- "We just need to get this over the line."
- "We can revisit that after launch."
- "It is not ideal, but it is good enough for now."
- "The MVP is close enough that cleanup can wait."

You might see:

- Workarounds become normal.
- Cleanup items keep moving forward with the release plan.
- Temporary implementation choices stop being marked as temporary.
- Each compromise makes the next compromise easier to justify.
- The product requires increasing care, local knowledge, or manual correction to keep moving.

Ask:

- What is still genuinely temporary, and what has quietly become the standard?
- What are we calling temporary that later work now depends on?
- Which compromise has become part of the product's shape?
- What condition would trigger revisiting this decision?

- Has that condition already arrived?
- What would it cost to reverse this now?

For the fuller pattern, see **Chapter 14 — Incremental Compromise.**

Process Erosion

You might hear:

- "It is just for the demo."
- "Do not let process slow us down."
- "We can document it later."
- "Just go straight to the person who can do it."
- "We do not need a ticket for everything."
- "The process is there, but this is urgent."

You might see:

- Important decisions happen in private messages, direct founder requests, or after-hours work.
- Formal reviews happen after the practical decision has already been made.
- The team keeps updating the process to match what already happened.
- Documentation, tracking, or signoff remains visible but no longer carries authority.
- The lived workflow routes around the official workflow whenever pressure rises.

Ask:

- Where does work really move when pressure rises?

- Is the difference between official process and lived process deliberate and visible, or informal and hidden?
- Can the process still interrupt momentum?
- Can it surface an unresolved question before the demo?
- Can a new person understand how work actually moves without a private translation layer?

For the fuller pattern, see **Chapter 15 — Process Erosion**.

Incentive Drift

You might hear:

- "This will look good for investors."
- "We need something impressive to show."
- "Everyone is talking about this space."
- "That feature will make the story easier to sell."
- "Reliability matters, but it will not move the room."
- "We need visible progress."

You might see:

- Reliability, testing, documentation, onboarding, and cleanup always lose to visible features.
- Work that photographs well, demos well, or reports well gets protected first.
- The same people are praised for speed while quieter structural work is treated as delay.
- Teams learn which questions reduce momentum and stop asking them early.
- Product direction starts serving attention rather than structural health.

Ask:

- What work gets energy before it has to argue for itself?
- What necessary work keeps losing because it is harder to admire?
- Does this solve the right problem?
- Does the system support it?
- Can customers actually use it?
- Does the reward signal still point toward the outcome the company claims to want?

For the fuller pattern, see **Chapter 16 — Incentive Drift.**

The Human Cost

Human Debt Accumulation

You might hear:

- "The team handled the last push."
- "They always find a way."
- "We are all busy; this is startup life."
- "Everyone is stretched right now."
- "That person knows the system best, so give it to them."
- "The senior team can absorb this for now."

You might see:

- Senior people spend increasing amounts of time rescuing, explaining, checking, or translating.
- New people increase coordination load rather than reducing pressure.

- Key people become routers for context, reassurance, and technical judgement.
- Knowledge lives in individuals rather than in the system.
- The team looks staffed, but one or two people are still carrying the real continuity.

Ask:

- What is the system quietly asking people to absorb?
- Who keeps being interrupted, explaining, remembering, smoothing tension, or carrying context?
- Where does the plan assume capacity that only exists because someone keeps compensating?
- What structural gap is being carried through helpfulness?
- Where has care become the substitute for structure?

For the fuller pattern, see **Chapter 19 — Human Debt Accumulation.**

Normalised Overextension

You might hear:

- "This is just what startup intensity looks like."
- "We need one more big push."
- "Everyone is tired, but that is normal right now."
- "We can recover after launch."
- "They did it last time, so they can probably do it again."
- "We do not have the luxury of slowing down."

You might see:

- The same people stay late, absorb interruptions, and carry the difficult context.

- Recovery disappears after major pushes.
- Approved leave becomes porous because the system depends on specific people.
- Plans assume effort levels that were previously treated as exceptional.
- Exhaustion is discussed as a personal condition rather than a system signal.

Ask:

- Does ordinary progress now depend on extraordinary stretch?
- Is recovery treated as part of delivery, or as a luxury people can take later?
- What did people survive last time that the next plan now assumes?
- When new work is added, what work is removed or deferred?
- Is endurance being treated as evidence of capacity?

For the fuller pattern, see **Chapter 20 — Normalised Overextension.**

Emotional Withdrawal

You might hear:

- "No one raised that in the meeting."
- "Everyone seemed aligned."
- "The room was calm."
- "The real concerns came up afterwards."
- "People are being professional."
- "The team is quieter, but that probably means they are focused."

You might see:

- Meetings are calm, but the real conversation happens afterwards.
- Risks live in code comments, side conversations, or private chats instead of formal tracking.
- People ask sharper questions in one-on-ones than in group settings.
- Concerns are raised as passing comments rather than full arguments.
- The strongest truth travels later, softer, or through narrower channels.

Ask:

- Do people still feel safe bringing the full weight of their expertise?
- Where does the truth go after the meeting?
- Are people still speaking, but withholding the sharper version of what they know?
- What concern is easier to raise in a side conversation than in the formal room?
- Has self-editing become part of how people survive the meeting?

For the fuller pattern, see **Chapter 21 — Emotional Withdrawal.**

The Reckoning

The Bill Comes Due

You might hear:

- "Why is this suddenly so expensive?"
- "How did this become a rebuild?"
- "Why are we only hearing about this now?"
- "This was supposed to be cleanup."
- "Why does every change need so much explanation?"
- "We have spent months on this and still do not know what it will take."
- "This is not a technical debt issue anymore; this is a business issue."

You might see:

- A technical review, audit, or report creates shock rather than relief.
- Problems that used to appear as small items now require months, funding, disruption, or executive attention.
- Estimates keep growing while confidence keeps shrinking.
- The same repair conversation returns with a larger cost attached.
- The organisation starts negotiating with constraints it previously treated as future cleanup.

Ask:

- What cost are we still treating as scattered after it has started becoming payable?
- What would have to be true for this to become ordinary cleanup again?

- Which repair keeps returning with a larger estimate?
- What deferred decision now needs funding, staffing, time, or executive attention?
- What becomes impossible if the bill is not paid?

For the fuller pattern, see **Chapter 24 — The Bill Comes Due.**

Compression of Consequence

You might hear:

- "This has come out of nowhere."
- "Why did nobody tell us?"
- "Customers are finding this before we can contain it."
- "We do not have weeks to work this out anymore."
- "The issue used to stay inside the team; now it is reaching the customer."
- "Every decision seems to create an immediate consequence."
- "We cannot absorb this quietly anymore."

You might see:

- Customers, leaders, adjacent teams, or the market begin encountering problems the team has known about for months.
- Issues that once stayed local now create visible consequences quickly.
- The organisation has less time between signal and response.
- People pre-filter ideas before meetings because consequences feel immediate.
- The operating rhythm shifts from learning to containment.

Ask:

- Where have we been relying on delay to make reality feel manageable?
- Who has been living with this reality longer than leadership has been recognising it?
- What used to stay local but now reaches customers, leaders, adjacent teams, or the market quickly?
- Which issue now lands in days instead of weeks?
- Are we mistaking urgency for clarity?

For the fuller pattern, see **Chapter 25 — Compression of Consequence.**

The Narrowing of Options

You might hear:

- "We still have several options."
- "We can always rebuild later."
- "Nothing has been decided yet."
- "We can pivot if we need to."
- "That option is technically possible."
- "Let's keep the door open."

You might see:

- The company discusses options that have no owner, resources, or willingness to pay the cost.
- Ideas are abandoned before discussion because everyone understands the constraints.
- Meetings shift from exploration to mitigation.

- Migration, simplification, or rebuild remains possible in theory but unavailable in practice.
- The language of choice remains wider than the field of action.

Ask:

- Which options are still real, and which are only being named?
- Does this option have an owner, resources, and a cost the organisation is prepared to face?
- What has waiting changed?
- Has waiting made the clean path more expensive, the simple path less credible, or the repair path more disruptive?
- Are we confusing possibility with availability?

For the fuller pattern, see **Chapter 26 — The Narrowing of Options.**

Forced Clarity

You might hear:

- "We need one source of truth."
- "We need to know who owns this."
- "Which number is right?"
- "What does the system actually do?"
- "What would a proper audit show?"
- "If we mapped this end to end, what would we find?"
- "We cannot keep carrying this in people's heads."

You might see:

- Reporting, ownership, product behaviour, customer data, or technical reality cannot be reconciled cleanly.

- People discover that different teams have been using different truths.
- A review exposes that the system depends on undocumented memory, partial data, or informal interpretation.
- Visibility feels threatening before anyone has acted on it.
- The organisation treats clarity as attack because clarity would force ownership.

Ask:

- What ambiguity is still carrying us?
- If this became visible end to end, what would it force us to own?
- Who owns this?
- Which source of truth wins?
- What must be corrected?
- What do we stop pretending is still fine?
- Is visibility being treated as threat or as responsibility?

For the fuller pattern, see **Chapter 27 — Forced Clarity.**

After Using This Index

Once this index has helped you locate the likely pattern, **Chapter 29 — Acting on Recognition** offers guidance on how to respond before deciding what to act on. It covers the difference between orientation and intervention, how to test whether a signal is real, and why a reset can look like learning without being it.

Cross-Pattern Starting Points

Use this section when the situation feels familiar but the pattern is not yet clear. These entries are starting points, not final diagnoses.

When everything takes more effort than it should

You might notice:

- Small changes require unusual care.
- Work that used to feel simple now needs context, explanation, and protection.

Begin with **Chapter 14 — Incremental Compromise.**

When the product keeps feeling close but not durable

You might notice:

- The product has looked close for a long time.
- The MVP exists, but cannot become durable.

Begin with **Chapter 10 — The Inevitability Illusion**, then **Chapter 14 — Incremental Compromise.**

When controlled success fails in the field

You might notice:

- The product works in the office, demo path, or reference environment, but struggles in the field.
- Controlled success is being used to dismiss deployment concerns.

Begin with **Chapter 9 — Narrative Lock-In**, **Chapter 10 — The Inevitability Illusion**, and **Chapter 25 — Compression of Consequence.**

When no one fully understands the whole thing

You might notice:

- Handover produces fragments rather than a coherent map.
- Screenshots, old reports, documents, repositories, and explanations do not agree with one another.

Begin with **Chapter 27 — Forced Clarity.**

When the room is calm but thinner than it should be

You might notice:

- The team is capable, but strangely quiet.
- The meeting feels calm, but the real truth appears afterwards.

Begin with **Chapter 21 — Emotional Withdrawal.**

When people are working harder but the company is learning less

You might notice:

- Senior builders are spending more time in meetings, coordination, and explanation than in construction.
- New headcount has increased communication load faster than it has increased clarity.

Begin with **Chapter 19 — Human Debt Accumulation** and **Chapter 20 — Normalised Overextension.**

When cleanup has become reckoning

You might notice:

- The problem suddenly needs months, money, or a rebuild.
- The work is no longer cleanup; it is repair, recovery, or reckoning.
- A cost that was previously scattered has condensed into a concrete obligation.

Begin with **Chapter 24 — The Bill Comes Due.**

When the promise is running ahead of the system

You might notice:

- Customer-facing certainty is increasing while delivery confidence is decreasing.
- More of the team is working around the mismatch between what was sold and what exists.

Begin with **Chapter 16 — Incentive Drift**, **Chapter 24 — The Bill Comes Due**, and **Chapter 25 — Compression of Consequence.**

When the stack, vendor, or platform dominates the conversation

You might notice:

- Technical choices that once looked fast or serious now shape what the company can realistically do.
- Reversibility exists in theory but keeps being deferred.

Begin with **Chapter 11 — Borrowed Confidence** and **Chapter 26 — The Narrowing of Options.**

When process has increased but clarity has not

You might notice:

- The company has added process, rituals, or delivery mechanics, but core confusion remains.
- Meetings and ceremonies have increased without increasing shared understanding.

Begin with **Chapter 15 — Process Erosion.**

When the company wants a fresh start more than it wants to understand what happened

You might notice:

- A reset feels emotionally cleaner than repair.
- The next version is being asked to solve problems the organisation has not yet named.
- The stack, platform, rebuild, or fresh start has become the visible explanation for deeper product, process, or leadership problems.
- Lessons are being named without changing the behaviour that made the reset feel necessary.
- Before the reset begins, no one can clearly answer what will change that makes the old pattern harder to reproduce.

Begin **with Chapter 9 — Narrative Lock-In** and **Chapter 14 — Incremental Compromise**.

When Several Signals Appear At Once

Do not treat the index as a hunt for one correct label. These patterns reinforce one another.

A story stops changing. Early progress begins to feel inevitable. Confidence is borrowed from an impressive person, partner, or tool. Temporary compromises accumulate. Process loses authority. Incentives reward the visible work. People absorb the difference. Exceptional effort becomes normal. Truth travels less freely. Costs become concrete, consequences arrive faster, options narrow, and clarity is eventually forced.

When several entries in this index feel familiar, begin with the earliest pattern you can still influence.

Ask:

- What story is shaping how we interpret the evidence?
- What are we calling temporary that has already become structural?
- Who is quietly compensating for the system?
- What consequence is getting harder to defer?
- Which option is still real, rather than merely discussable?

The purpose of diagnosis is not to prove that the company is failing. It is to restore enough clarity that the next decision is made from reality rather than from the pattern currently filtering it.

Reflection Questions by Part

This appendix brings together the central recognition questions from each part of the book so they can be used as a working tool rather than encountered only at the end of individual chapters.

These questions are for founders, leadership teams, boards, product leaders, and senior technical staff who want to examine the condition of a startup before the signals become consequences. They are not meant to produce quick agreement. Their value comes from slowing the room down long enough for different altitudes, experiences, and interpretations to become visible.

Use one section at a time. You do not need to answer every question; begin with the two or three that feel most uncomfortable. Ask each person to answer independently before discussing the questions as a group. Pay particular attention to questions that produce very different answers across roles. That difference is often where the most useful signal begins.

Part I – Reading the Weather

Part I asks whether the organisation can see its own condition clearly before trying to correct it.

- What are people closest to the work experiencing that is not visible in leadership reporting?
- Where are summaries, dashboards, milestones, or demos compressing important detail?
- Which signals are being treated as local problems even though they recur across teams or systems?
- What feels different inside the company even though the formal story sounds stable?

- Where has belief stopped being useful fuel and started becoming a filter?
- Which explanations have become so familiar that they are no longer being tested?
- What do different levels of the organisation believe is happening right now?
- Where do those interpretations diverge most sharply?
- What would become uncomfortable if the company described its present condition without optimism, blame, or reassurance?
- Which signals are already visible but do not yet feel important enough to interrupt momentum?

Do not begin by asking what needs fixing. Begin by asking whether the organisation is reading the same weather. If people at different altitudes are describing materially different conditions, the first task is not intervention. It is restoring enough shared visibility that later decisions are based on the same reality.

Part II – The Allusions

Part II examines the stories, signals, and sources of confidence shaping how the startup interprets what it sees.

- What story is the company currently telling itself about the product, the market, the team, or the technical approach?
- When did that story last change because of evidence?
- Which concerns keep returning without materially changing the plan?
- What has been demonstrated, what has been proven, and what has merely become easier to believe?
- Where is visible progress carrying more certainty than it has earned?

- Which future capabilities are being spoken about as if they already exist in unfinished form?
- What confidence comes from the current system, and what confidence is being borrowed from a person, vendor, brand, platform, tool, investor, or prior success?
- Are external partners helping the company see reality more clearly, or making reality feel safer than it is?
- Which assumptions would be hardest to reopen because too much identity, credibility, or effort now depends on them?
- What evidence would genuinely cause the company to revise its story?

The purpose of this discussion is not to weaken belief. It is to distinguish earned confidence from narrative momentum. A useful startup story remains open to correction. Once every new signal is interpreted as confirmation, the story is no longer guiding learning. It is governing perception.

Part III – The Accumulation

Part III asks what reasonable decisions have started compounding into structure.

- What is still being described as temporary even though later work now depends on it?
- Which compromises have been carried forward without a clear condition for revisiting them?
- Where does the product require increasing care, local knowledge, or controlled conditions just to preserve ordinary progress?
- What work keeps returning with a larger cost, wider impact, or more fragile workaround?

- Where does the formal process differ from the way work actually moves under pressure?
- Which decisions occur through side channels, direct requests, or exceptions that bypass shared visibility?
- Can the process still slow a weak decision, expose a risk, or force unresolved work back into view?
- What work receives energy automatically because it is visible, impressive, fundable, or easy to report?
- What necessary work keeps losing because it is harder to admire?
- Are incentives still pointing toward the outcome the company claims to want?

Look for sequences rather than isolated decisions. One shortcut, exception, or misplaced priority may mean little. Repetition is the signal. Accumulation begins when the organisation keeps making locally sensible choices that collectively reduce trust, clarity, reversibility, and shared understanding. The central question is not whether compromises exist. It is whether the company still knows what they are becoming.

Part IV – The Human Cost

Part IV examines where unresolved structural strain is being stored in people.

- Who repeatedly explains, rescues, checks, translates, remembers, reassures, or catches problems before they become visible?
- Which people are functioning as the system's undocumented memory or extra margin?
- Where does the staffing plan assume capacity that exists only because senior people keep compensating?

- What forms of invisible support work are required to make visible delivery appear successful?
- Has exceptional effort become the assumed baseline for ordinary progress?
- Where has recovery disappeared from the operating model?
- Which late nights, interruptions, outbursts, or signs of depletion have been treated as personal incidents rather than system signals?
- Are concerns being raised fully in formal meetings, or only afterwards in smaller, safer conversations?
- Do people still believe that bringing the full truth into the room will produce useful movement?
- What is the company quietly asking people to absorb that should instead be resolved structurally?

This discussion is not an audit of individual resilience. The point is to identify where the organisation is financing continuity with human attention, patience, trust, judgement, health, and recovery. When people keep making the system work, leaders can easily conclude that the system is working. These are not the same thing.

Part V – The Reckoning

Part V asks what can no longer be deferred, distributed, or kept ambiguous.

- Which costs have stopped behaving like cleanup and started requiring months, funding, staffing, disruption, or executive ownership?
- What problem keeps returning with a larger estimate and less certainty of recovery?
- Which consequences are now reaching customers, leaders, or adjacent teams faster than they once did?

- Who has been living with the underlying reality longer than leadership has been recognising it?
- Which options are still genuinely actionable, and which are only being named in discussion?
- What would each remaining option require in money, time, trust, authority, and organisational energy?
- Which choices became unavailable because earlier decisions made them too expensive, disruptive, or politically difficult?
- What ambiguity is still allowing the company to postpone ownership?
- If the system became visible end to end, what would the organisation be forced to admit or decide?
- Is new visibility being treated as useful evidence, or as a threat to the people and stories already invested in the current path?

The reckoning is not automatically the end of the company. It is the end of a particular form of distance from consequence. The leadership task is to name what is now real, separate repairable cost from unrecoverable loss, and stop discussing constrained choices as though the strategic field remains wide open.

Part VI – After Recognition

Part VI turns recognition into a more deliberate response without confusing visible action with learning.

- Is the signal recurring across story, accumulation, human load, and consequence, or is it still an isolated event?
- What are we observing directly, and what meaning are we adding to it?
- What would make this pattern harder to reproduce, rather than merely easier to manage temporarily?

- Are we continuing because the path remains sound, or because stopping would expose sunk cost and identity?
- Are we restarting because we have learned, or because a reset allows us to avoid the conditions that produced the first result?
- Where does the next commitment become harder to reverse?
- What smaller intervention could interrupt the pattern before a larger correction becomes necessary?
- How can the concern be stated in observable language rather than accusation?
- What signal can people without final authority preserve, document, or keep from being softened away?
- What would a better outcome look like if rescue, scale, and preservation of the original story were not the only measures of success?

The aim is not to produce a heroic intervention. It is to attach effort to reality. Sometimes the right response is repair. Sometimes it is narrower scope, a slower commitment, a simpler technical path, clearer ownership, a boundary, restructuring, or an earlier and cleaner ending. Recognition creates leverage when it changes what the organisation is willing to see before deciding what it is willing to do.

Reading the Patterns from Governance Altitude

Most of this book is written from inside the company. That choice matters because the patterns described here are easiest to feel from inside the work: in the product decisions, delivery pressure, technical trade-offs, founder conversations, team fatigue, informal workarounds, and small signals that appear long before the company has language for them.

Boards and investors sit at a different altitude.

They are close enough to influence direction, capital, confidence, leadership behaviour, and strategic choices. They are also far enough from the daily work that the company often reaches them through compressed forms: board papers, dashboards, milestone updates, forecasts, demos, founder narratives, and selected conversations with senior leaders.

That distance is not a flaw. It is part of the role. A board cannot operate at implementation depth, and investors cannot inspect every system they fund. Their value comes from judgement, pattern recognition, challenge, perspective, and the ability to see beyond the immediate pressure consuming the operating team.

But the distance creates a specific risk.

The board may not be seeing the company. It may be seeing the company's explanation of itself.

Compressed Reality

The central governance problem is not that boards receive incomplete information. Boards always will.

The problem is that incomplete information can arrive in a form that feels complete.

A board pack may include revenue, runway, delivery status, hiring, risks, customer activity, product progress, and strategic priorities. Each item may be accurate. The distortion enters through selection, compression, interpretation, and tone.

The people preparing the material are usually trying to hold several legitimate concerns at once. They need to report honestly. They need to preserve confidence. They need to avoid drowning the board in implementation detail. They need to protect the team from unnecessary interference. They need to show that management remains in control.

None of that requires dishonesty, but it can produce a picture in which the facts are true while the condition is still misread.

A product may be described as delayed when the deeper issue is that the product boundary remains unclear. A release may be amber when the real condition is dependence on repeated rescue work. Customer interest may be reported as validation when it is closer to curiosity, goodwill, or a controlled-path success. Technical debt may appear as future cleanup when it has already narrowed strategic options.

The board sees the dashboard.

The operating team feels the weather.

Good governance depends on narrowing that gap without collapsing the board into operational management.

Three Common Misreads

Boards and investors are especially exposed to three misreads that run through this book.

The first is **momentum as validation**.

New customers, pilots, hires, releases, product demonstrations, funding conversations, partnerships, and visible activity all create the impression that the company is advancing. The question is not whether those things matter. They do. The question is what they have proved.

A successful demo may prove that a controlled path works. It may not prove repeatability, operational readiness, supportability, customer adoption, data integrity, or product-market fit. A signed pilot may prove interest. It may not prove durable demand. A growing team may prove commitment to scale. It may not prove that the product boundary, architecture, or operating model can carry that scale.

The second misread is **credibility as proof**.

A respected executive joins. A known consultancy is engaged. A major technology platform is adopted. An experienced founder has succeeded before. A credible investor is on the cap table.

All of those signals matter. None of them replaces evidence from the current system.

Borrowed confidence becomes a governance risk when reputation reduces scrutiny before the company has earned trust in its own operating conditions. The useful question is not only whether the person, partner, tool, or prior success is credible. It is whether their involvement is increasing the company's ability to see and own reality.

The third misread is **calm as alignment**.

A board meeting can become smoother as the company becomes less healthy.

Cleaner updates, fewer disagreements, and more consistent messaging may indicate maturity. They may also indicate that difficult signals are being softened before they arrive. People close to the work may not stop speaking because they have changed their view. They may stop because experience has taught them that the formal system does not use the full truth well.

By the time visible conflict reaches the board, the organisation may have spent months reducing the intensity of the signal.

The task is not to manufacture conflict. It is to test whether challenge still travels.

The Board Shapes the Signal

Boards and investors do not only receive information. They shape the conditions under which information is produced.

Founders and executives learn quickly what the board rewards. If the strongest response goes to confidence, green status, visible growth, and decisive answers, management will naturally optimise the way information is framed. That does not require deception. It requires adaptation.

If uncertainty is repeatedly treated as lack of control, uncertainty will be compressed before it reaches the board. If technical reality is treated as operational detail, technical reality will be translated into thinner summaries. If changing direction is interpreted as weak conviction, the story may be protected longer than the evidence supports.

The board's reactions become part of the incentive system.

This is why the quality of board questions matters. Questions do not only extract information. They teach the company what kind of truth is welcome.

The most useful board questions keep optimism answerable to evidence. They make room for uncertainty without rewarding vagueness. They distinguish between management not yet knowing and management refusing to look. They ask whether progress has been demonstrated, proven, or merely made easier to believe.

Questions Worth Asking

These questions are not a substitute for ordinary governance. They are a way to read beneath the update.

- What has been demonstrated, and what has been proven under real operating conditions?
- Which part of the current confidence comes from repeatable evidence?

- What only works because people close to the system know how to make it work?
- Where is the company borrowing confidence from a founder, senior hire, vendor, investor, partner, platform, tool, or prior success?
- What assumption is carrying the most strategic weight right now?
- When did that assumption last change because of evidence?
- What concern has appeared repeatedly without materially changing the plan?
- Where do senior leaders and people closest to the work interpret the same condition differently?
- What information reaches the board only after it becomes an incident?
- Which options are genuinely actionable rather than merely discussable?
- What are we treating as reversible that has already become difficult to unwind?
- How might our own questions, funding decisions, reactions, or appetite for certainty be shaping what management feels able to say?

The purpose of these questions is not to make boards more operational.

It is to keep the signal alive at governance altitude.

From The Board Seat

Boards and investors cannot see everything. They should not try to.

But they can notice when a company is becoming less able to see itself.

They can notice when the story stops changing, when visible progress carries more certainty than it has earned, when external credibility substitutes for internal clarity, when the strongest concerns arrive late or softened, and when the same people keep absorbing the difference between the company as described and the company as lived.

Governance does not sit outside the patterns described in this book. It participates in them.

The board helps shape what is rewarded, what is spoken, what is delayed, and what becomes difficult to admit. Investors shape which futures feel fundable and which forms of progress receive legitimacy. Their influence can widen the company's capacity to see reality, or increase the pressure to keep presenting a reality that remains investable.

The best boards do not remove uncertainty. They improve the organisation's ability to hold uncertainty without turning it into distortion.

From the board seat, the work is not to see everything.

It is to notice when the company is becoming less able to see itself.

Additional Resources

The Founders' Folly is written as a book of recognition: a way to see recurring startup patterns before they become expensive.

Some readers will want to take the ideas further into live conversations with founders, leadership teams, boards, investors, or technical groups. I maintain additional resources for that purpose, including discussion guides, board and investor questions, diagnostic prompts, and current writing on how these patterns appear in changing startup conditions.

These materials help teams use the book without turning it into a rigid framework or checklist. Their purpose is the same as the book's purpose: to make important signals easier to recognise while there is still room to respond.

You can find current resources at thefoundersfolly.com/resources

About the Author

Gary Smart has spent three decades working in and around early-stage and growing technology businesses, often close to the pressure points where product ambition, technical reality, and organisational strain begin to pull against each other. Across software engineering, greenfield product work, advisory, and leadership-adjacent roles, he has worked alongside founders, technical leaders, and senior decision-makers whose choices shaped company trajectory long before the consequences were obvious.

Over time, that experience gave him an unusually close view of how intelligent teams, strong products, and compelling vision can still drift into predictable patterns. His work now centres on helping founders, technical leaders, and other high-influence decision-makers see those patterns earlier: where belief becomes a filter, where compromise is quietly accumulating, and where human load is carrying more than the system should.

The Founders' Folly draws on that lived experience. It is an attempt to name the structural patterns that shape startup trajectories before they are misread as isolated mistakes, personality flaws, or sudden bad luck.

For articles, workshops, and resources related to *The Founders' Folly*, visit garysmart.com.au or connect with Gary on LinkedIn.

Website:	garysmart.com.au
LinkedIn:	linkedin.com/in/gnsmart
Email:	hello@garysmart.com.au